Introduction to Forestry

Introduction to Forestry

Lucas Gordon

MURPHY & MOORE
www.murphy-moorepublishing.com

Murphy & Moore Publishing,
1 Rockefeller Plaza,
New York City, NY 10020, USA

Visit us on the World Wide Web at:
www.murphy-moorepublishing.com

ISBN: 978-1-63987-331-9 (Hardback)

Cataloging-in-Publication Data

Introduction to forestry / Lucas Gordon.
p. cm.
Includes bibliographical references and index.
ISBN 978-1-63987-331-9
1. Forests and forestry. 2. Forest management. 3. Forestry innovations.
I. Gordon, Lucas.
SD131 .I58 2022
634.9--dc23

Table of Contents

Preface VII

Chapter 1 **Forestry: An Introduction** **1**
- Agroforestry 16
- Energy Forestry 26
- Sustainable Forestry 36
- Ecoforestry 39
- Wildlife Forestry 41
- Social Forestry 51

Chapter 2 **Forest: Conservation and Management** **56**
- Types of Forest 60
- Forest Conservation 61
- Forest Management 64

Chapter 3 **Afforestation** **76**
- Reforestation 89
- Urban Forestry 96
- Biosequestration 102
- Climate Change Mitigation 124

Chapter 4 **Deforestation** **159**
- Illegal Logging 181
- Deforestation and Climate Change 187
- Forest Dieback 189

Chapter 5 **Products Derived from Forest Resources** **194**
- Woodworking 194
- Paperboard 198
- Fiberboard 203

- Wood Veneer 204
- Wood Preservation 207
- Timber Recycling 226
- Non-Timber Forest Product 228

Permissions

Index

Preface

The science of managing, conserving and utilizing forested land and its associated wasteland and waterways primarily for the purpose of obtaining timber and fuel is called forestry. It also focuses on conservation of woodlands and other resources associated with forests for environmental and human benefits. Forestry is widely practiced in plantations and natural stands. It is a knowledge based practice of managing forest land for protection, recreation, commercial and industrial use as well. Forestry is an inter-disciplinary field which involves the principles of biology, physics, social and managerial science which deals with a wide variety of concerns related to multiple-use management. Water quality management, landscape and community protection, erosion control, employment, preservation of forests as sinks for atmospheric carbon dioxide, and biodiversity management are some key areas of focus under this discipline. This book elucidates the concepts and innovative models around prospective developments with respect to forestry. It is compiled in such a manner, that it will provide in-depth knowledge about the theory and practice of forestry. The book is appropriate for those seeking detailed information in this area.

A detailed account of the significant topics covered in this book is provided below:

Chapter 1- Forestry is the practice of managing and maintaining forested lands, together with associated water and wasteland, for commercial, public and agricultural uses. Agroforestry, bamboo forestry, energy forestry, sustainable forestry and ecology forestry are a few forms of forests which fall under its domain. This chapter serves as an introduction to the significant aspects of forestry.

Chapter 2- The practice of planning and maintaining forested areas for the sustainability of future generations is known as forest conservation. The branch of forestry deals with the administrative, social, economic, scientific and technical aspects is referred to as forest management. This chapter provides comprehensive information regarding forests and their conservation and management methods.

Chapter 3- Afforestation can be defined as the act or process of planting trees with the ultimate aim of establishing a forest on land not previously forested. Reforestation refers to the process of replanting trees in forested and woodland areas which have undergone depletion. Urban forestry, biosequestration and climate change mitigation are other topics related to afforestation which are discussed at length in this chapter.

Chapter 4- Deforestation is the large scale removal of trees in a forest or woodland area to provide clearings for other human activities such as agricultural croplands, urbanization or mining activities. It has brought about consequences like soil erosion, climate change, biodiversity depletion, etc. Deforestation and its associated features like illegal logging, forest dieback, climate change, etc. are covered in detail in this chapter.

Chapter 5- The resources obtained from forests are finite and have to be utilized judiciously. The integration of forest ecology and biology, forest measurements, forest policies and administration to manage the resources extracted from forest is called forest resource management. This chapter deals with forest products and forest resource management in an extensive manner for a better understanding of the subjects.

I would like to make a special mention of my publisher who considered me worthy of this opportunity and also supported me throughout the process. I would also like to thank the editing team at the back-end who extended their help whenever required.

Lucas Gordon

Chapter 1

Forestry: An Introduction

Forestry is the practice of managing and maintaining forested lands, together with associated water and wasteland, for commercial, public and agricultural uses. Agroforestry, bamboo forestry, energy forestry, sustainable forestry and ecology forestry are a few forms of forests which fall under its domain. This chapter serves as an introduction to the significant aspects of forestry.

Forestry is the art and science of managing forests so as to yield, on a continuous basis, a maximum in quality and quantity of forest products and services. In broad sense, it is for the handling of forest land to satisfy the needs of man. It includes the logging, manufacturing, marketing and use of wood products. Forests can be managed for single or multiple purposes, to include protection of watersheds, production of timber, wildlife conservation, and regulation of stream flow, recreation, erosion control and general aesthetics.

Forest ecosystems have come to be seen as the most important component of the biosphere and forestry has emerged as a vital field of science, natural resource management, applied art and technology. The main goal of forestry is to create and implement socio-environmental systems that allow forests to continue a sustainable continuation of environmental supplies and services. The challenge of forestry is to create systems that are socially accepted while sustaining the resource and any other resources that might be affected.

Branches of Forestry

- Agroforestry: It is the integration of forests into agricultural systems in order to optimize the production and positive effects within the system and minimize negative side effects of farming.
- Dendrology: It involves the study and identification of economically useful tree species.
- Forest Ecology: It studies the patterns and processes of a forest ecosystem.
- Forest Economics: It studies the impact of economics on forest management decisions.
- Forest Hydrology: It embodies the effects of changes in forest land use on the movement, distribution, and quality of water in the ecosystem.
- Forest Management: It comprises the overall administrative, economic, legal and social aspects of forest regulation.
- Forest Mensuration: It incorporates quantitative measurements of the forest stand to

determine stand timber volume and productivity/health and provides a basis of which management decisions can be made.

- Forest Protection: It comprises both the biotic and abiotic maladies affecting the health of the forest or tree, primarily by fungal pathogens, insects, nematodes, parasitic plants and climatic factors (stress, salinity, fire, etc.) and their management.
- Silviculture: It is the art and science of controlling the establishment, growth, composition, health and quality of forests to meet specific objectives.
- Tree Breeding: It is the method of genetically modifying/selecting forest stock for improved growth or vigor characteristics.
- Urban Forestry: It entails the care and management of urban tree populations for the purpose of improving the urban environment.
- Social Forestry: It addresses human-forest interactions and the importance of community based natural resource management.
- Farm Forestry: The practice of cultivating and managing trees in compact blocks on agricultural lands.
- Community Forestry: Combination of forest conservation with rural development and poverty reduction objectives, accomplished through instating a legal framework that favours profitable and sustainable forest management.
- Eco-Forestry or Environmental Forestry: It emphasizes tree planting practices which strive to protect and restore ecosystems.
- Energy Forestry: It includes specifically managing forests for the production of energy from biomass or bio-fuel derived from a fast-growing species of tree or woody shrub.
- Short Rotation Forestry: Managing a forest that utilizes fast-growing species as a bio-based energy crop for domestic use and commercial use in power stations, alone or in combination with other fuels such as coal.
- Analog Forestry: A management focus that seeks to establish a tree-dominated ecosystem that is similar in architectural structure and ecological function to the naturally occurring climax and sub-climax vegetation community.
- Boreal Forestry: It analyzes the particular challenges of forestry in the world's boreal regions.

Rotation Recognized in Forestry

The period which a forest crop takes between its formation and final felling is known as rotation or production period. Generally, rotation is expressed as an average age practice for harvesting to start a new crop.

This term is also defined in various other ways by different authors and some of these definitions are given below:

- The planned number of years between the formation or regeneration of a crop and its final felling. In the case of a selection forest, the average age at which a tree is considered mature for felling.
- The number of years fixed by the working plan between the formation or regeneration and the final felling of a crop.
- Rotation age is the age of trees or crops at which they are felled and object of management for the time being are best served.
- Rotation or production period is the interval of time between the formation of a young crop by seeding, planting or other means and its final harvesting.
- Rotation is the period which elapses between the formation of a wood and the time when it is finally cut over.

Types of Rotation

Rotation is an important factor in the regulation of yield and proper management of the forest as a whole. As stated earlier, it will depend mainly on the objects of management. Various types of rotation recognized in forestry are as follows:

Physical Rotation

It is the rotation which coincides with the natural lease of life of a species on a given site. The natural life-span of trees varies greatly with species and the site factors. This rotation is applicable only in case of protection and amenity forests, park lands and in some cases roadside avenues.

It is very variable, fairly long and also indefinite. Another interpretation of physical rotation is the age up to which the trees remain sound, or produce viable seed in high forests and, in coppice crops, can put forth reliable coppice shoots. This rotation is not of any relevance to economic forestry.

Silvicultural Rotation

It is the rotation through which a species retains satisfactory vigour of growth and reproduction on a given site. It can neither be lower than the age at which trees start producing fertile seed in sufficient quantity, nor beyond the age when they stop doing so. It is also necessary that soil conditions remain satisfactory for germination and establishment of seed. It is not only long but has also very wide range of limits, hence somewhat vague and may be used in combination with other rotations, such as Technical rotation.

Technical Rotation

It is the rotation under which a species yields the maximum material of a specified size or suitability for economic conversion or for special use. It aims at producing the maximum material of specific dimension for specific purpose, such as railway sleepers, saw-logs, mine-props, transmission poles, match-wood, paper-wood etc. Since one and the same tree may yield different assortments of material and the trees in a crop may attain given size at different times, the technical exploitable age offers no reliable fixed point for fixing the rotation.

It does, however, allow for fixation of limits within which a tree or stand is adopted for the production of assortments in greater number or better quality. Technical rotation is adopted particularly by industrial firms which own forests for the purpose of supplying raw material for their plants (e.g., NEPA and West-coast paper mills, WIMCO match factory, ITC, etc.).

Rotation of Maximum Volume Production

It is the rotation that yields the maximum annual quantity of material i.e. the age at which the Mean Annual Increment (M.A.I.) culminates. The M.A.I, refers to that of the stand and not that of individual trees. More often than not, the objective in forestry is quality or value production and rotation for higher value production is usually longer than for highest volume production. In turn, maximum volume rotation is usually longer than financial rotation. Common practice in forestry is to adopt a combination of rotation of maximum volume production and financial rotation.

Rotation of Highest Income

It is the rotation which yields the highest average annual gross or net revenue irrespective of the capital value of the forest. It is calculated without interest and irrespective of the times when the items of income or expenditure occur. This rotation is important from the overall national point of view.

With forestry in the public sector, attainment of highest gross revenue is more important than that of net income because larger expenditure and investment generates several social benefits and indirect advantages to the trade and industry. The private owner of a forest estate is interested in maximum net revenue by keeping the rotation period as short as possible.

The average net annual revenue or rental obtained from a stand of trees is expressed by the formula:

$$\text{Mean annual net revenue per unit area} = \left(Yr + \sum Tr - C - \sum E\right)/R$$

where,

Yr = Value of final felling per unit area.

Tr = Value of all thinning during rotation period 'R' per unit area.

C = Cost of formation of stand per unit area.

E = Annual cost of administration per unit area.

R = Rotation (years).

The two rotations will be about the same unless there is an appreciable increase in price for larger-sized timber which is, in fact, usual. If the size gradient is marked, then the rotation of highest net income will be comparatively longer. Again, if there is a special size of timber which fetches a particularly high price, the rotation which provides that price may be the rotation of highest income and possibly coincides with technical rotation.

Financial Rotation

It is the rotation which yields the highest net return on the invested capital. It differs from the rotation of highest net income in that all items of revenue and expenditure are calculated with compound interest at an assumed rate, usually the rate at which the Govt., is able to borrow money. It is also defined as:

- The rotation which gives the highest discounted profit, usually at its commencement.
- The rotation which is most profitable.
- The rotation which gives the highest net return on capital value i.e. under which the soil expectation value calculated with a given rate of interest is the maximum.

Rotation ages of some important plantation species are as follows:

- Teak – 60- 120 yrs,
- Eucalypts, Poplars and Casuarinas – 4 – 8 yrs,
- Acacias – 8 – 12 yrs,
- Prosopis – 4 – 8 yrs.

Length of Rotation

The choice of the type of rotation will depend on the object of management, but the length of rotation of whatever type, will depend on the interaction of several physical and economic factors given below:

- Rate of Growth: This will vary with species, site fertility, intensity of thinning, etc.
- Silvicultural Characteristics of the Species: For example, natural span of life, age of fertile seed production, age at which rate of growth culminates, age at which the quality of its timber is most desirable or begins to fall, etc.
- Response of the Soil: Deterioration / exhaustion of soil by exposure, biotic influence etc.
- Economic Considerations: Depending on a combination of factors of cost, prices of different sizes, time required to reach those sizes, etc. A factor tending to lengthen the rotation is the increase in value of large-sized timber; though not economically sound to grow large-sized timber over a long rotation.
- Social Conditions: Socio-economic and employment policy of the State.

Costs on Forestry

In most valuation studies, while the benefits from forestry have been well documented, the costs incurred on its cultivation, maintenance, restoration, depreciation and other miscellaneous costs to obtain the timber, non-timber forest products, eco-tourism and other benefits are often not accounted for.

Here, opportunity costs are also ignored. Restoration costs are incurred for reproducing the original level of benefits. Certain expenditures are required to prevent degradation of the forests. Further, input costs are incurred on plantation, extraction cost, marketing expenses for forestry products, especially log wood and other NTFPs.

Direct expenditure on afforestation includes land preparation, pitting, planting, watering, fencing, trenching, apart from purchase of seedlings, fertilizers, after care cost, etc. Apart from the direct costs, there are certain indirect costs that are required to be incurred in order to reap the larger benefits associated with forests.

Some of these costs would include infrastructure cost for promoting eco-tourism, markets for timber and NTFPs, etc. It is, however, important to note that most of these are in the nature of joint costs and attributing these costs exclusively to forestry activities could be misleading.

Most of the investment/expenditure in forests is undertaken by public agencies. Hence it is also important to look at the opportunity cost of land under forests or proposed for afforestation. This would require estimation of opportunity cost of alternative uses of the land such as grazing, agriculture, horticulture, habitation, industry, etc. The opportunity cost would reflect all economic outputs foregone or precluded by maintaining land under forest cover.

Maintenance of existing natural forests would imply prospective benefits foregone from agriculture or other alternative uses of land. Thus, opportunity cost of sustainable forest management is the amount that could be earned from forest exploitation and various forms of land utilization. In case of converting wasteland under the afforestation programmes, the opportunity cost could be in terms of restricted availability of grazing and pasture land, etc.

Opportunity cost of benefits from forest would also indicate the shadow price of benefits arising from watershed catchment protection and soil conservation, nutrient cycling, maintenance of soil fertility amongst others. The other issue relates to social costs that are associated with existence of forests on the local people.

Such social costs could take the form of damage to resilience of forest eco-system, natural resource depletion, pollution on account of forest fires, etc. Another widespread cost results from crop damage caused to local farmers by animals and birds dwelling in nearby forests.

Application of Remote Sensing and GIS in Forestry

The forest is a complex ecosystem consisting mainly of trees that have formed a buffer for the earth to protect life forms. The trees which make up the main area of the forest create a special environment which, in turn, affects the kinds of animals and plants that can exist in the forest. In the tropical and subtropical region, forests are further subdivided into plantations and natural forests.

Natural forests are forests composed of indigenous trees, not deliberately planted. Plantations are forests established by planting or/and seeding in the process of afforestation or reforestation.

Efficient and sustainable management of India's forests therefore, is critical, given the significance of biomass to rural communities. Materially, forests provide fuel and fodder, food, materials for construction, and agricultural implements, but crucially, forests also provide income-generating opportunities, supplying materials that can be employed in cottage industries such as bamboo, leaves, lac, honey, herbs and fruits etc. for direct sale. Using remote sensing, foresters can get more accurate and cost-effective information and can directly observe as large an area as necessary.

Due to the versatility and scale of remote sensing, it is invaluable in all stages of forest management. The forester's task begins with growing healthy forests. Remote sensing is a useful tool for assessment of environmental conditions, either in an existing forest or prior to planting a new one. Effective management of forest requires detailed local information on the status of forest. However, the traditional method of ground based surveys to obtain such detailed inventory data is currently expensive, time consuming and difficult to put into practice.

Remote sensing and Geographic Information System (GIS) analyses have great potential for providing relatively cheap data to assist in the efficient management of forests. The Global Positioning System (GPS) technology is now being used extensively for field inventory.

Forestry and Life Cycle Assessment

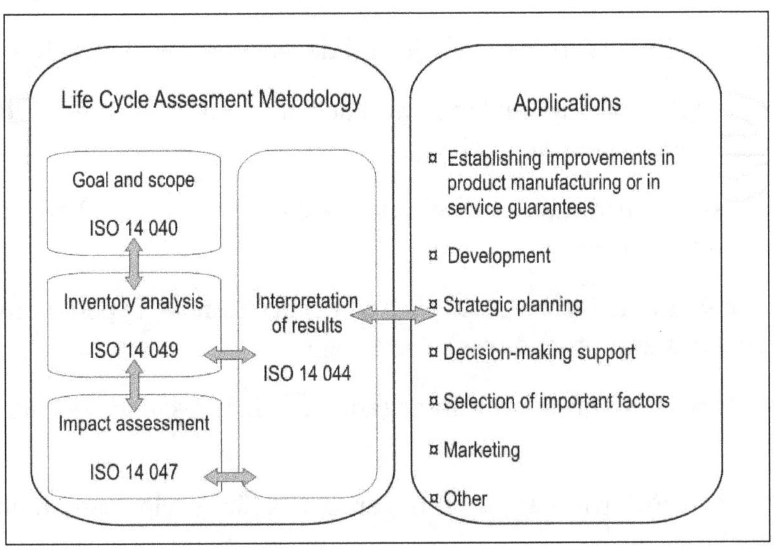

Fig: Phases in the LCA according to ISO standards.

Ever since recognizing the effect of global warming and ozone degradation, scientists are trying to develop new methods of environmental impact assessment to ensure environmental protection. Life Cycle Assessment (LCA) is such a method and with its fully comprehensive approach better nature conservation is possible. Sustainability can be achieved through minimal consumption of renewable and non-renewable materials, energy saving, reuse and recycling, emission control, etc. However, the cost aspect is still the most important factor in today's world and therefore achieving sustainability is made even more difficult. LCA according to FAO is a useful tool for comparing the environmental aspects of specific products as it enables the ecological comparison of two or more products made of different raw materials but used for the same purpose. LCA can help in measuring environmental aspects and potential impacts of a product through its entire life-cycle from raw material acquisition (as the beginning of the life-cycle) to manufacturing, use, recycling, reuse and final disposal (as the end of the product's life-cycle). The ISO/EN 14040 defines LCA as a technique for assessing the environmental aspects and potential impacts associated with a product by: 1) compiling an inventory of relevant inputs and outputs of a system; 2) evaluating potential environmental impacts associated with those inputs and outputs; 3) interpreting the results of the inventory analyses and 4) impact assessment in relation to the objectives of the study.

Goal and scope of the LCA explain why the study is being done, set functional units (for example 1 m3 or 1 m2 or 1 tonne are usually taken for wood products) and give system boundaries to determine which unit-processes of LCA will be considered in the study (data quality requirement, precision and representatives of data). Inventory analysis indicates relevant input data (used energy, raw and supporting materials etc.) and output data (releases to air, water, land and manufacture of side products, by-products etc.). To avoid high complexity of the inventory analysis, processing machines, manpower, buildings and land-use for biological production or transportation are not included. Impact assessment reveals the potential environmental repercussions and their significance. Impact categories can be defined form local to global. Interpretation of results combines all the findings together and final conclusions and recommendations can be made.

Other ISO/EN standards deal with LCA as well:

- ISO 14020: 2000 (Environmental labels and declarations – General principles).
- ISO 14021: 1999 (Environmental labels and declarations – Self-declared environmental claims, Type II environmental labeling).
- ISO 14024: 1999 (Environmental labels and declarations – Type I environmental labelling – Principles and procedures).
- ISO 14025: 2006 (Environmental labels and declarations – Type III environmental declarations – Principles and procedures).
- ISO 14040: 2006 (Environmental management – Life cycle assessment – Principles and framework).
- ISO 14044: 2006 (Environmental management – Life cycle assessment – Requirements and guidelines).
- ISO/TR 14047: 2003 (Environmental management – Life cycle impact assessment – Examples of application of ISO 14042).
- ISO/TS 14048: 2002 (Environmental management – Life cycle assessment – Data documentation format).
- ISO/TR 14049: 2000 (Environmental management – Life cycle assessment – Examples of application of ISO 14041 to goal and scope definition and inventory analysis).

According to COST ACTION E9 Memorandum of Understanding (Life cycle assessment of forestry and forest products) Europe currently produces and consumes the roundwood equivalent of 400 million m3 of wood products. Forests moreover cover more than 25% of the world's land area with approximately 600 billion m3 of standing stock.

Annually 3.6 billion m3 of wood is used as firewood (55%) or as different products (45%). Because wood is a renewable energy source, CO_2 neutral and recyclable, it represents the most important renewable raw material and fuel and a very important carbon sink. Even more importantly, wood can be permanently available if sustainable forest management is made obligatory. The main reasons of performing LCAs by COST Action E9 of forestry and forest products are:

- To enable comparison between different materials, provided that products are used for the same purpose.
- To obtain quantified and reliable information of benefits of wood products and their environmental impact, so that industry and policy makers can use such information.
- To highlight unknown or uncertain areas of environmental impact of wood products.
- To improve production and recycling techniques by minimizing steps of high environmental impacts or choosing different processing routes to reduce such high environmental impact.

However, certain problems when dealing with LCA in forestry arise:

- Forestry uses considerable areas of land.
- Life cycle of forest products can vary from relatively short (for example paper) to very long (for example structural timber).
- Long production chains which start in the forest and end with disposal or burning for energy.
- Forest products have complex relationships between products, by-products and waste.

Complexity of the LCA study is shown in figure below.

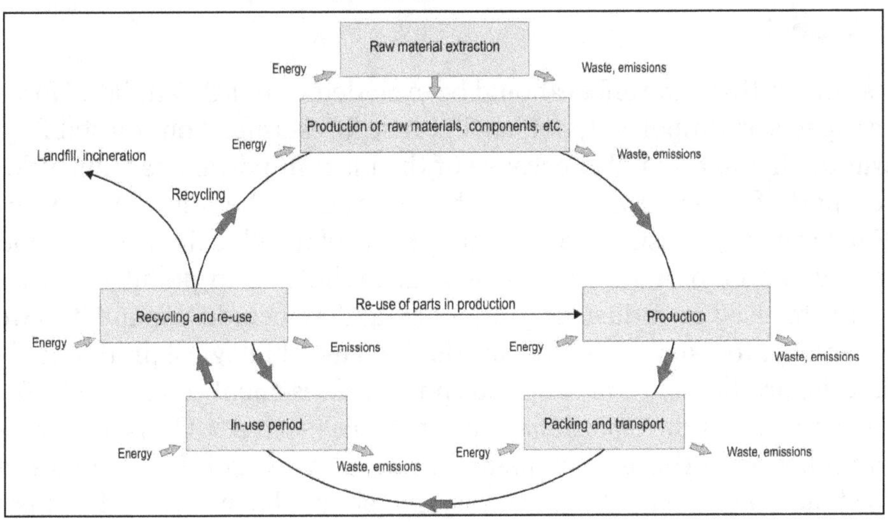

Fig: Product lifetime.

Challenges for LCA in Forestry

The carbon cycle starts with the biosynthesis in the forest and ends by releasing CO2 into the atmosphere during combustion or biodegradation. The most important material and energy cycles of wood as a renewable raw material should be therefore considered in LCA studies. Production of timber requires large area of land, but different kinds of land use should be also considered and compared, for example timber production in forests in comparison to agriculture raw material production, renewable forest material and fuel production in comparison to the exploitation of non-renewable materials and fuels and of land use in industry. Different forest management systems should also be compared and studied. Complexity of LCA studies for forestry and relating industries is shown in Figure below.

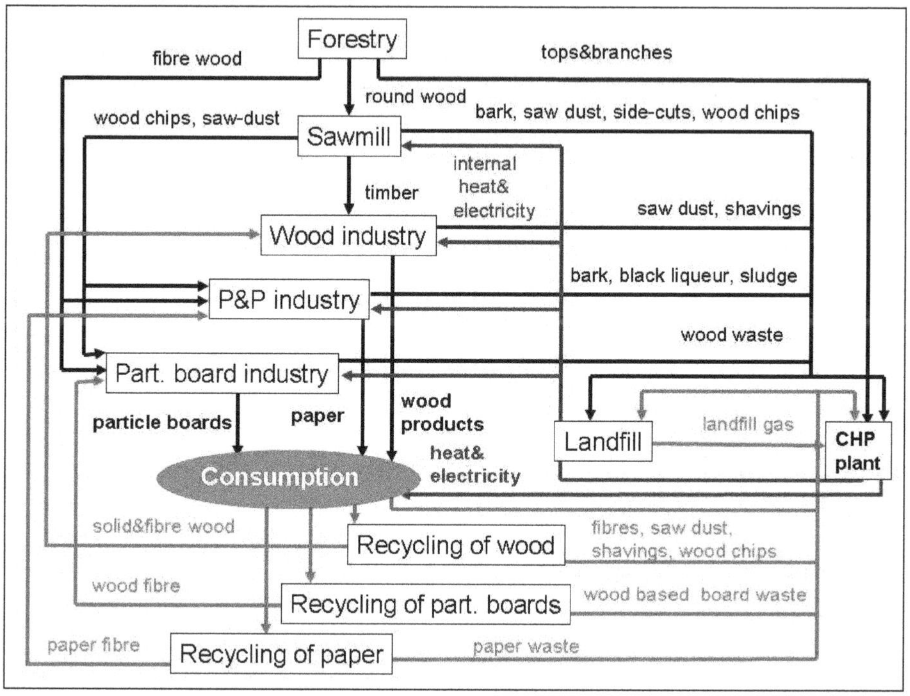

Fig: Complexity of forestry products by Jungmeier.

Space-time Aspect

Many scientists believe that time effect should be excluded from LCA studies. However, in forestry social and time aspects are important, not just because forests effect on soil stability, water quality, biodiversity, wildlife habitat, but also because of the most important carbon cycle. It emphasizes that forests are a part of the environment and therefore should be considered as an impact category of its own. Furthermore, forests are a substantial part of the global ecosystem and are negatively influenced not only by poor or inadequate management, but also by polluted air, acid rain etc. LCA is a method mainly focused on industrial production and to include/expand it to wood production and make it possible for forestry it should consider land use (biological production) and site conditions of raw material production. Only a serious impact assessment can characterize a substance as an environmental burden. If the emissions of a substance (such as CO_2) is interpreted as a burden, consequently a reduction of this environmental burden by fixing this substance in the biological growth process should be stated in the inventory and should be interpreted as a benefit.

Even though in industrial production it is sufficient to concentrate on flows between different phases of a product lifetime, in forestry that is insufficient and stocks should also be considered. To make this possible, raw material acquisition i.e. forest production should be standardized. Standard argument for exclusion of beneficial effects forests have on the environment is a lack of methods for measuring such effects.

To understand and quantify the role of the forest ecosystem in the carbon cycle it is necessary to quantify both the net annual carbon fluxes (Mg C ha-1 yr-1) and the total carbon.

Content (Mg C ha-1) of representative forest ecosystems, thereby including the carbon fluxes and stocks in the soil. Virtanen and Nilsson made a life cycle analysis of waste paper recycling, where CO_2 fixation was 0.7 to 0.9 kg for each kg of biomass (logs including bark and water content). Calculations of net CO_2 uptake for 1 kg of biomass were 0.46 kg CO_2-equivalents, which gave 0.92 kg CO_2-equivalents per kilo of usable wood (assuming 50% of water content). Grasser makes CO_2 uptake calculations in a stack-wood chain. The chain stars with wood formation, and the CO_2 uptake is based on a carbon content of 0.5 kg per kilo of oven dry matter, which means 1.833 kg of CO_2 uptake per kilo of usable wood (including bark). To provide a kilo of usable wood (including bark), 1.2 kg of wood needs to be cut, of which 0.2 kg will stay in the forest as wood waste and decay and the bounded CO_2 within will be set free again during its biological decomposition. This represents 17% of the total growth. Lox) calculates CO_2 uptake of produced cardboards and states that biological decomposition of wood waste if different. 34% of the total growth will stay in forest, which means 0.5 kg of wood waste per kilo of usable wood (including bark) because only cellulose and hemi cellulose fraction will decompose, which means 2.2 kg of CO_2 uptake per kilo of usable wood. Different ways to calculate the CO_2 uptakes have been mentioned and De Feyter states that the CO_2 should not be seen as a credit, but as an implementation of the carbon neutrality of wood when looking at its lifecycle. The emitted carbon can be present in the form of CO_2, CO, CH_4 etc. and those emissions have a different behaviour towards environmental effects which can result in different CO_2-equivalents.

When comparing recycling and non-recycling of paper in LCA, most analysis show a higher contribution to global warming of recycling systems. This is because recycling systems use more fossil fuels, however recycling of paper holds carbon longer in fixed rotation, but this aspect cannot be taken into account in LCA because LCA still has a static approach and no time aspects.

Wollenman investigates aims at balancing carbon sequestration and emission fluxes of forest land use to evaluate methodology for a close-to-nature and plantation regime. The regimes have a positive sequestration capacity of about 85 kg C m3 (eucalyptus plantation) and about 180 kg C m3 (close-to-nature beech forest) respectively. Assuming a given area of forest cover a division of so called ecological labour between plantation forest and forest reserve provides a net carbon sink capacity of 0.18 kg C m-2a-1 compared to 0.14 kg C m-2a-1 for close- to-nature regime. These results clearly indicate that management regime determines the net carbon sink capacity and that so called ecological labour division between intensively and unmanaged forests offer opportunities to maximize carbon sequestration.

Forest roads provide accessibility to forests, and their construction is one of the largest investments in forestry itself. Multi-purpose use of forest roads creates problems involving the process of building and maintaining forest roads in the life cycle assessment of forest products. Lifetime of roads is difficult to define, since it depends on many factors and generalizes their depreciation over time. Mroueh et al in a study of life cycle assessment of road construction analyze numerous factors (environmental loadings, Table) that need attention:

- Consumption of energy and fuels (including energy consumed by the machine, or the vehicle during the processing of raw materials and energy contained in organic materials).
- Emissions and dust.
- Environmental noise pollution.
- Substances released into the soil depending on the type of building material (for example sulphate compounds, mercury, arsenic, chromium, copper, vanadium, etc.).

Table: Environmental loadings in the life cycle assessment of road construction.

Impact category	Environmental loading	Results, unit
Resource use	Use of natural resources	t/construction selected
	Industrial by products	t/construction selected
	Energy	kWh/construction
	Fuels	m3/construction
	Land use	Verbal estimation of the significance of land use
Effluents to soil and waters	Leaching of metals (e.g. As, Cd, Cr, Cu, Mo, Ni, Se, Pb, Zn)	mg/m^2 of construction selected
	Leaching or migration of compounds from materials	mg/m^2 of construction selected
	CI, SO_4	mg/m^2 of construction selected
Emissions to air	CO_2	kg/construction selected
	NO_X	
	SO_2	
	VOC	
	CO	
	Particles	
Wastes	Inert waste	t/km
Other loadings	Noise	Noise x time/km

- Only constructions that meet the same performance requirements and are designed for the same site are compared in the assessment.
- Fuel means diesel oil used to power machines and vehicles or oil used as a raw material in industry. Can be presented separately or included in the energy consumption.
- The substance included in the analysis is selected according to the material used.
- The frequency of repairs is determined in advance adopted strategy.
- Water-soluble substances present in the materials can be carried away by run-off water

into the environment and into the groundwater. The amount of leaching depends on the composition of the material, the amount of water passing through material and the manner in which it is laid. Pollutants leaching out of alternative constructions over a hundred years are presented in table below.

Some of the most important factors, such as COD emissions to water and land, have been absent from this study and applicable calculation methods requires more extensive further research. The creation of procedures for calculating the environmental loadings would be important for assessing the impacts of material selections and the total loadings of road usage.

Table: Noise levels of working machines.

Machine type	Noise level range	Average noise level
	dBA	
Drilling rig	98 – 101	100
Blasting	125 – 136	130
Hydraulic hammer	87 – 92	90
Conveyor belt	84	84
Crushing plant	100	100
Hydraulic excavator	82 – 100	89
Earth moving machines	91	91
Lorry	84	84
Bulldozer	80 – 89	84
Road roller	84– 101	92
Asphalt layer	74 – 89	81
Road grader	85 – 89	87

At a distance of 7 m from the source.

Table: Environmental loadings caused by road construction and maintenance.

Environmental loadings	Construction	Maintenance
CO2, kg/km	263 000 – 562 000	33 900
SO2, kg/km	280 – 610	4,1
NOX, kg/km	2600 – 3800	140
CO, kg/km	600 – 1100	20
Violatile organic compounds (VOC), kg/km	550 – 980	210
Fuel consumption, l/km	63 000 – 100 000	18 200
Energy consumption, kWh/km	790 000 – 1 .470 000	183 300

The classic LCA includes a static approach that ignores the dimension of time and flows are determined through emissions that occur in a certain production period, while the duration of the period is not taken into account. Forestry on the other hand just by using dynamic models predicts the production and thus the time dimension is completely necessary for modelling. Furthermore, the accumulation of substances in the classical analysis of life cycles are ignored and only the flows of goods between production systems and the environment are taken into account, while in forestry supplies represent the basic information for modelling. It should be noted that the static analysis of the life cycle for forestry has not yet been performed because of the mentioned dynamic character of the forestry production systems in general. in the short term analysis it is easier to implement dynamic models in the study of forestry especially in view of the circulation of substances (CO_2, NH_3, NO_3, etc.).

Table: Pollutants leaching out over hundred years.

Substance	Ash FA1	Ash FA2/FA3	Concrete CC1	Concrete CC2	B-F slag BFS
	mg/m2				
Sulphate	692000	446000	546000	791000	-
Chloride	84600	44600	13800	20800	-
Arsenic	84	69	-	-	<0.1
Barium	7.7	7.7	-	-	-
Cadmium	0.06	0.05	-	-	0.4
Chrome	250	140	92	131	<0.1
Copper	4.4	2.7	51	65	<1
Mercury	0.3	0.2	-	-	-
Molybdenum	1260	1260	-	-	<0.1
Nickel	19	12	-	-	<0.1
Lead	0.15	0.15	-	-	4
Selenium	45	28	-	-	-
Vanadium	615	615	-	-	5
Zinc	3	2	-	-	<10

Domino effect caused by land use begins with the influence of vegetation on the soil and ground, the influence of the nutrients and water in the soil, soil compaction (the probability of erosion, and thus the loss of fertile soil), until the end of the product life cycle, through which one can see the total impact of products on the environment.

Cowell and Clift proposed how to solve the problem of soil erosion in the life cycle assessment. They believe that soil washout is a category of impact, the consequence is the erosion of fertile soil (kg/ha/yr), and indication of the erosion is presented with global supplies of fertile soil. Global supplies of fertile soil are determined as the ratio of total upper layer of soil in the world (tonnes) and of an annual loss of the upper layers of soil due to erosion (t/yr).

Forest Products

The definition of waste in the context of life cycle analysis is given in the standard ISO 14040 ("output from the production process that no longer can be used"), but in forestry waste can get a new meaning. A typical example of waste transferring to product in forestry is in the extraction of forest residue for energy purposes (branches and tree tops for example which are a by-product of wood in the timber assortment production, and usually represent waste in wood cuttings). In case the forest residue is left to decompose in the forest, i.e. additional profit will not be made; it represents a loss or a waste.

As well as the term "waste" is more complex when dealing with forestry, complications arise in use of other terms for products. A group of standards, ISO 14000, gives definitions for a main product (a product or a co-product which depends on economics and on the conversion of production process itself), a final product (product which requires no additional transformations prior to its use), a by-product (the product is not the intended output of a process and there is a market for the product; economic value is relatively low), a co-product (any of two or more products from the same unit process; production of this product is the intended output of the process with a defined market and economic value is relatively high) and a waste (any output from the production system which is disposed of and will not go further processing). It is clearly defined that a product is a physical things with measurable characteristics such as size, colour, taste, etc., resulting in the production process or services offered to consumers and requested by the consumer. Definitions of other sub-products depend entirely on the current market conditions. In wood processing many "waste streams" are used for internal heat or energy, which can also cause problems in the allocation procedure and definition of system boundaries. Wood itself, as an output, can be referred to as "kg absolute dry" or in "kg with X% moisture content dry base".

Apart from wood products there are many other products which can be gained from the forest and also have market value. Up till 2001 and COST E9 2001 workshop (Energy, carbon and other material flows in the life cycle assessment of forestry and forest products) the non- wood products have not been mentioned in the LCAs on wood products and no environmental impacts have as yet been allocated. Compared to the mass of wood produced and harvested in the forests, the amount of non-wood products is generally bellow 1%. Made a survey of reported non-wood products in tonnes per year according to the national data-bases and UN- ECE/FAO study.

Table: Reported non-wood products in tonne per year.

State	Resin	Mushrooms	Game and other hunted animal t/yr	Berries	Reindeer (Rangi-fertarandus)	Cork	Chestnuts	Others
Austria	-	-	-	-	-	-	-	-
Denmark	-	-	-	-	-	-	-	-
Finland	-	1408	5857	8441	2200	-	-	311
France	-	8200	-	1000	-	4000	-	600
Germany	-	-	-	-	-	-	-	-
Greece	6140	-	-	-	-	-	-	-
Ireland	-	-	-	-	-	-	-	-
Italy	-	2614	-	496	-	9204	69852	22658
Norway	-	1200	7787	25000	-	-	-	1500
Slovenia	-	800	840	600	-	-	-	3000
Sweden	-	8500	17000	20700	-	-	-	26163
Switzerland	-	735	1597	-	-	-	-	525

Finland uses 30.1 million m3 of peat for energy and 1.6 million m3 for horticultural and bedding peat out of forests.

Agroforestry

Agroforestry, the combined cultivation of tree species and agricultural crops, is an ancient and still widespread practice throughout the world. It encompasses a variety of land use practices and systems.

In agroforestry systems, woody and herbaceous perennials are grown on land that also supports agricultural crops or animals. The mixture of these components, in the form of spatial arrangement or temporal sequence, enhances ecological stability and production sustainability. This integration allows the components to complement one another in their use of resources and in the timing of that use. Perennials have deeper roots and higher canopies than those of annuals, allowing better management of above- and below-ground resources. Under ideal conditions:

- Nutrients recycled from the subsoil to the surface by deep-rooted perennials can be used by annuals.
- Leguminous perennials fix atmospheric nitrogen that can be used by annuals.
- There is minimal competition for water because of differences in depth from which the roots of annuals and perennials extract water from the soil.
- Some perennials produce allopathic compounds that can suppress weeds.
- Differences in the structure of perennials and annuals, leading to a multistory canopy, reduce competition for light among plants.

Agroforestry systems have the potential to improve production and to enhance the agronomic and ecological sustainability of resource-poor farmers in the humid tropics. In practice, however, the potential benefits of agroforestry systems can be harnessed only through skillful and labor-intensive management of compatible systems. There are no simple blueprints of a universally applicable system that can harness all potential benefits possible under ideal conditions. Thus, a wide range of agroforestry systems has been designed to alleviate agronomic, ecologic, or managerial constraints.

Types of Traditional Agroforestry Systems in the Humid Tropics

Agroforestry is not a new concept in the humid tropics. Several types of traditional agroforestry systems exist, but no standard classification system is available to categorize them. Nair proposed a classification system based on structural, functional, agroecological, and socioeconomic factors (Figure). These broad categories are interrelated, and not necessarily mutually exclusive. In agroforestry land use systems, three basic components are managed by people: the tree (woody perennial), the herb (agricultural crops, including pasture species), and the animal.

Based on their structure and function, agroforestry systems can be classified into the following three categories:

- Agrisilviculture is the use of crops and trees, including shrubs or vines. It includes shifting cultivation, forest gardens, multipurpose trees and shrubs on farmland, alley cropping, and windbreaks as well as integrated multistory mixtures of plantation crops.
- Silvopastoral systems are combinations of pastures (with or without animals) and trees. They include cut-and-carry fodder production, living fences of fodder trees and hedges, and trees and shrub grown on pastureland.

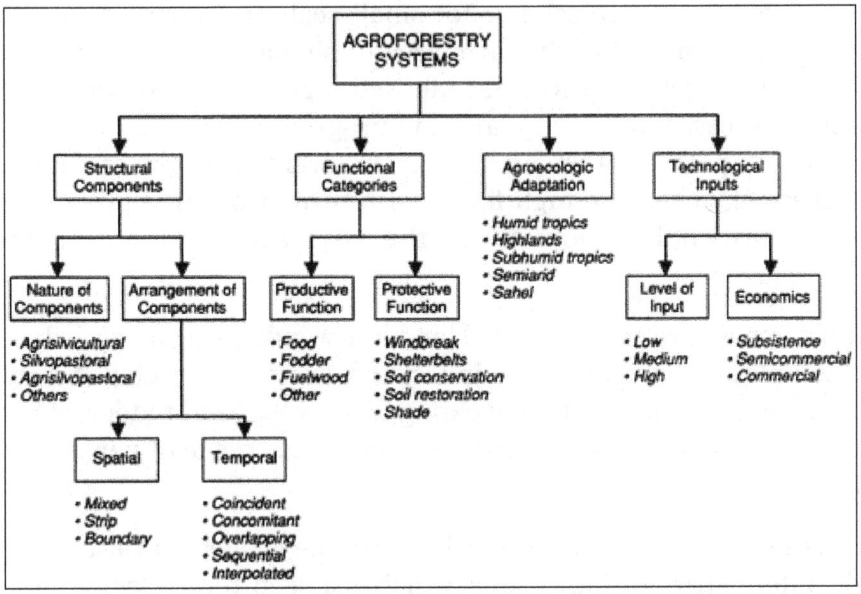

Fig: Characteristics of traditional agroforestry systems used in the humid tropics.

Agrisilvopastoral systems are those that combine food crops, pastures (with or without animals), and trees and include home gardens and woody hedges used to provide browse, mulch, green manure, erosion control, and riverbank stabilization.

Other types of agroforestry systems include apiculture (beekeeping) using honey-producing trees, aquaculture whereby trees lining fishponds provide leaves as forage for fish, and multipurpose woodlots that serve various purposes such as wood, fodder, or food production and soil protection or reclamation.

Principal types of agrisilvicultural systems traditionally used in the humid tropics are:

- Rotational agroforestry: In traditional shifting cultivation, trees and wood species are naturally regenerated over a period of 5 to 40 years and rotated with annual crops that are cultivated from 1 to years. Improved tree species can be grown in place of native vegetation to achieve better soil conditions. This technique is used in multipurpose woodlots (where diverse mixtures of trees are used), home gardens (where trees and crops are grown close to the house), and compound farms (where trees, animals, crops, and the farmer's dwelling are in a fenced area).
- An intercropping system: Annual and perennial groups of plants are grown within the

same land management unit. This system enables continuous production of food and tree products with a minimum need for restorative or idle fallow. Typical examples of intercropping systems include alley cropping and boundary planting of trees and wood hedges.

Two examples of the successful use of agroforestry systems by resource-poor farmers in the tropics are found in the Philippines and Rwanda. In the Philippines, many small-scale farmers took up cash-crop-tree farming to produce pulpwood, poles, timber, charcoal, or fuelwood in the 1960s. The program gained significant momentum in 72 when the Paper Industries Corporation of the Philippines (PICOP) entered into an agreement with the Development Bank of the Philippines to develop a loan scheme for small-scale tree farmers with titled or untitled land. Provision was made for part of the farm area to be maintained under food crops. PICOP guaranteed a minimum purchase price, but allowed farmers to sell wood to other outlets if they could get better prices. Within 10 years, the program covered 22,000 ha and supported 3,800 farmers, about 30 percent of whom had taken advantage of the credit program. A key to the program's success has been the high financial returns from tree growing. Adequate market incentives and security of land tenure were the basic factors responsible for acceptance by farmers.

The second example involves restoring eroded land in Rwanda using an agrisilvopastoral system. At Nyabisindu, a complex system of trees, animals, and crops was developed using the community's existing knowledge. Trees and hedges, yielding fruit, wood, and fodder, were used as protective ground cover against soil erosion. Extensive use was also made of perennial crops to further stabilize the soil.

In Amazonian Ecuador, a sustainable system has been developed to raise sheep in association with cassava and contour strips of Inga edulis, which is a deep-rooted leguminous fuelwood tree. After the cassava is harvested, a perennial leguminous ground cover, Desmodium sp., is planted between the trees to enrich the soil. Sheep graze on the ground cover.

Keys to the success of these projects included building on traditional knowledge, involving farmers in the choice of species, and providing economic incentives greater than those of traditional systems. Resource conservation and land restoration were additional benefits to the local community.

The viability and sustainability of these systems can be attributed to some combination of the following factors:

- A reduced fallow period and a greater ability to cultivate on a long-term basis, thereby eliminating the need to move to new land.
- Reduced use of chemical fertilizers and other fossil-fuel-based inputs due to enhancement of soil organic matter and improvement in soil fertility.
- Improved soil structure and physical properties (for example, better sizes of pores and channels in the soil that allow better water penetration and drainage).
- Decreased risks of soil degradation from accelerated erosion and other degenerative processes.

- Increased production and a rise in economic status from subsistence to partially commercialized farm.
- Decreased need for clearing new land.

Improved Agroforestry Systems

Scientists and policymakers generally are eager to improve traditional agroforestry systems by enhancing productivity and ecological compatibility. Ways of improving these systems include using better trees and woody shrubs and creating an orderly arrangement of trees, crops, and livestock.

Improved Trees and Woody Shrubs

Several trees and woody shrubs are used in traditional and natural fallow systems. Some commonly used species include Acioa baterii, Afzelia bella, Alchornea cordifolia, Anthonotha macrophylla, and Gliricidia septum. Some improved species have several advantages in an agroforestry system, including their ability to fix nitrogen, grow fast, tolerate soil acidity, and withstand regular coppicing.

Multipurpose trees can also be grown on cropland or pastureland. They may be planted randomly or according to systematic patterns on embankments, terraces, or field boundaries. They provide a variety of products including fruit, forage, fuelwood, fodder, shade, and fence and timber material. Some commonly recommended multipurpose trees are listed in Once again, local adaptation to and validation for site-specific systems are essential.

Table: Net Primarly Production of Biomass for Commonly Recommanded Multipurpose Tree Species in the Humid Tropics.

Species	Net Primary Production of Biomass (Kg/ha/yr)
Acacia auriculiformis	3,000-4,000
Acacia mangium	2,500-3,500
Albizia falcate	4,000-5,000
Alchornea cordifolia	2,000-3,000
Calliandra calothyrsus	2,500-3,000
Cordia alliodora	2,500-3,500
Dalbergia latifolia	4,000-5,000
Erythrina poeppigiana	4,000-6,000
Gmelina arborea	1,500-5,000
Leucaena leucocephala	3,000-5,000

Arrangement of Trees, Crops and Livestock

Rather than using a random and difficult-to-mechanize system of growing trees with annuals or animals, mixtures can be grown in an improved spatial or temporal arrangement. In an agrisilvicultural system, for example, trees can be grown in alternate rows or strips, as contour hedges to control erosion, or on field boundaries. These orderly arrangements can facilitate the use of animal power and of mechanization of farm operations, save labor, and enhance economic and ecological benefits.

Alley cropping is a common example of a spatial arrangement. Food crops are grown in alleys formed by contour hedgerows of trees or shrubs. Trees and shrubs can be pruned to prevent shading of the food crops and to provide nitrogen-rich mulch for crops and fodder for livestock. Shrubs and trees also act as windbreaks, facilitate nutrient recycling, suppress weed growth, decrease runoff, and reduce soil erosion.

The most common trees for alley cropping are fast-growing, multipurpose, nitrogen-fixing trees. Tree species with the potential for use with nonacid tropical soils include Acioa baterii, Alchornea cordifolia, Gliricidia septum, and Leucaena leucocephala. Species for acid soils include Acioa baterii, Alchornea cordifolia, Anthonotha macrophylla, Calliandra calothyrsus, Cnestis ferruginea, Dialium guineense, Erythrina spp., Flemingia congesta, Harungana madagascariensis, Inga edulis, Nuclea latifolia, and Samanea saman. Hedgerows of Cassia spp., G. septum, and L. Ieucocephala can be established from seed. Other species are established from seedlings or stem cuttings. However, the use of stem cuttings often results in a patchy stand with a high rate of mortality. Trees established from stem cuttings are also easily uprooted because of poor root system development.

When successfully established, alley cropping systems can produce two or more products, such as food grains, fodder, mulch, fuelwood, and staking and building materials, and can increase or maintain soil structure. However, the beneficial effects of these systems depend on many factors, such as the tree species, area of land allocated to trees, hedgerow management, crop management, soil type, and prevalent climate. In areas with nonacid soils, satisfactory yields of cereals can be attained with the added benefit of erosion control. These systems are also labor intensive, therefore they are adapted primarily to areas of high population density and modest to low labor cost.

Advantages and Disadvantages of Agroforestry

Given a compatible association of trees and annual crops, agroforestry systems are likely to sustain economic productivity without causing severe degradation of the environment. Because of the low fertility of most upland tropical soils, some degradation is inevitable with any cultivation system. The rate and risks of such degradation are lower with agroforestry than with annual crop rotations. Soil organic matter, pH, soil structure, infiltration rate, cation exchange capacity, and the base saturation percentage are maintained at more favorable levels in agroforestry systems due to reduced losses to runoff and soil erosion, efficient nutrient recycling, biological nitrogen fixation by leguminous trees, favorable soil temperature regime, prevention of permanent changes in soil characteristics caused by drying, and improved drainage because of roots and other biochannels.

It is important to note, however, that trees have both positive and negative effects on soils. Negative effects include growth suppression caused by competition for limited resources (nutrients, water, and light) and by allelopathic effects. Mismanagement of trees (through, for example, improper fertilizer application or inadequate water control) can also cause soil erosion, nutrient depletion, water logging, drought stress, and soil compaction.

Economic evaluation is an important tool to assess a technology. Labor-intensive alley cropping can be economical under severe cash constraints and where hired labor is available at relatively low cost. The available data on alley cropping indicate that the system cannot sustain production without supplemental inputs of chemical fertilizers if high yields are desired. In fact, soil degradation and attendant yield reductions can occur even with the fertilizer application.

Erosion control is a definite advantage of closely spaced contour hedgerows of L. Ieucocephala or other shrubs, but it can also be achieved through cover crops, grass strips, or no tillage. None-theless, the erosion preventive effects of L. Ieucocephala hedgerows must also be considered in evaluating the economic impact of an alley cropping system.

Data on soil properties indicate that intensive cultivation resulted in decreases in soil organic matter content, total nitrogen, pH, and exchangeable calcium, magnesium, and potassium in all systems including alley cropping and control. This drastic decline in soil fertility was observed in relatively fertile soils (Alfisols). The relative rates of decline, however, were somewhat less in alley cropping than with prow-based control. These results are also supported by data on acidic tropical soils in Yurimaguas, Peru. Szott observed significantly more calcium, magnesium, phosphorus, and potassium in the upper 15 cm of soil with control without trees treatments than with alley-cropping treatments. Fertilized control without trees significantly exceeded all other treatments in topsoil calcium and magnesium. The pH values were also significantly greater in the fertilized control.

Classification of Agroforestry

Agroforestry system has been classified by the researchers on the basis of structural, functional and socio-economic aspects.

Structural Basis

The structural classification depends on different components of the agroforestry system and the individual role or function of each component. This type of classification is totally dependent on the type of components and their arrangement.

Nature of the Components

Agroforestry systems on the basis of nature of the components include:

- Agric-silviculture: Plantation of trees and crops.
- Agri-horticulture: Plantation of fruit trees and crops.
- Agri-silvi horticulture: Plantation of trees, fruit trees and crops.
- Agri-silvipasture: Plantation of trees, crops and pasture/animals.
- Alley cropping: Plantation of perennial hedges and crops.
- Aqua-forestry: Combination of trees and fishes.
- Boundary plantation: Plantation of trees on boundaries along with the crops.
- Block plantation: Block of trees and crops.
- Energy plantation: Combination of trees and crops during initial years.
- Silvi-olericuture: Plantation of trees and vegetables.
- Horti-olericulture: Plantation of fruit trees and vegetables.

- Horti-pasture: Combination of fruit trees and pasture/animals.
- Silvi-pasture: Combination of trees and pasture/animals.
- Shelter-belts: Combination of trees and crops.
- Forage forestry: Combination of forage trees and pastures.
- Wind –breaks: Plantation of trees and crops.
- Silvi or Horti-sericulture: Combination of trees or fruit trees and sericulture.
- Horti-apiculutre: Combination of fruit trees and honey bees.
- Aqua-forestry: Combination of trees and fishes.
- Homestead: Multiple combinations of fruit trees, trees, vegetables etc.

Arrangement of Components

Spatial arrangement: Agroforestry mixture that results in dense or sparse stands of plants is known as spatial arrangement. The species can be planted at varying widths or zones. Alley cropping (hedge row intercropping) represents common example of zonal pattern.

Temporal arrangement: Conventional shifting cultivation cycle is an extreme example of temporal arrangement. When a woody species is planted, then this type of arrangement involves 2-4 years of cropping and fifteen years of fallow cycle. Temporal arrangements are also termed as overlapping, coincident, concomitant, interpolated and separate.

Functional Basis of Classification

This classification is based on the major function of the woody components and it can be productive or protective. Productivity and sustainability are the fundamental attributes of the Agroforestry systems.

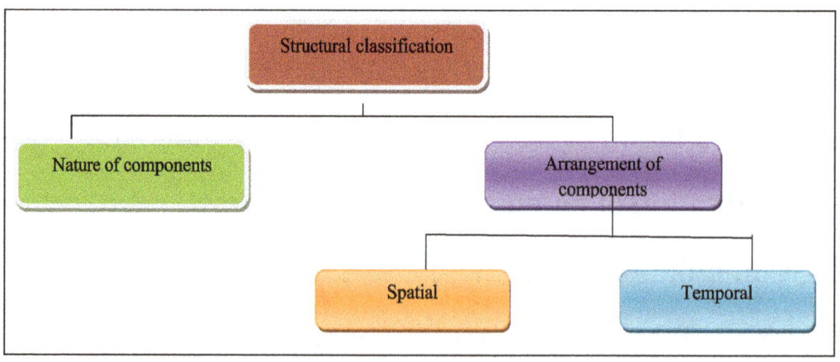

Fig: Diagrammatic representation of functional classification.

Productive functions of agroforestry system includes:

- Fodder.
- Fuel-wood.

- Gums, resins, waxes.
- Furniture wood.
- Other products.

Protective Functions

Protective functions of agroforestry systems include:

- Shelter-belt.
- Wind-break.
- Soil conservation.
- Soil improvement.
- Moisture conservation.
- Shade.

Socioeconomic Basis of Agroforestry Systems

This type of classification considers level of inputs (low or high), management,intensity/scale and commercial goals.

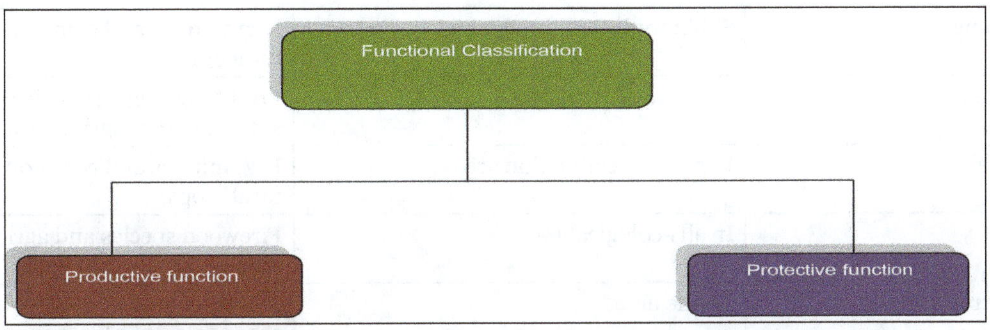

Fig: Diagrammatic representation of socio-economic classification.

Commercial System: Commercial Means Production of Particular Product on Large Scale

The commercialization process may be owned by government, private or corporate. Payment base of labour is either normal or contract base.

Examples include:

- Production of agricultural crops on commercial scale.
- Production of shade tolerant crop plants like tea, cacao and coffee.
- Commercial grazing under timber and pulp plantations.

Intermediate Agroforestry system: This system is known as an intermediate system since it comes in between commercial and subsistence scale of perennial cash crop production.

Examples: Plantation of cacao, coffee, fruit trees and timber species are included in intermediate agroforestry system. Subsistence agroforestry system: This system involves the land use for satisfying basic needs. This system is managed primarily by the owner and his family.

Examples:

- Production of cash crops.
- Traditional shifting cultivation.

Major Agroforestry Practices and their Agro Ecological Adaptability

Agroforestry practices have been categorized into three major classes:

- Agrisilvicultural.
- Silvi or silvopastural.
- Agrosilvopastural or agrisilvipastural.

Agroforestry practice	Agro ecological adaptability	Major components
Agrisilvicultural systems		
Alley cropping	Subhumid and humid areas	Leguminous and common agricultural crops.
Home gardens	All ecological regions	Fruit trees, vines, woody species and shade tolerant agricultural plants.
Improved fallow	In shifting cultivation areas	Leguminous and common agricultural crops.
Fuelwood	In all ecological regions	Firewood species and agricultural crops.
Multilayer tree gardens	Fertile areas	Woody components and sometimes-shade tolerant plants are also present. Herbaceous plants are usually absent.
Multipurpose trees on crop lands	In all ecological regions	Multipurpose trees, fruit trees and common agricultural crops.
Plantation crop combinations	Tropical humid/sub humid highlands or humid lowlands	Coffee, cacao, fruit trees, coconut, fodder, fuel wood and shade tolerant species.
Shelter breaks, wind breaks and live hedges	In wind-prone areas	Tall trees and agricultural crops.
Taungya	In all ecological regions	Forestry species and common agricultural crops.
Trees in soil reclamation and conservation	Sloping areas	Multipurpose tress, fruit trees and common agricultural species.
Silvopastural systems (trees in combination with pasture/animals)		

Protein banks	Areas with person: land ratio high	Leguminous and fodder trees.
Trees on range-land or pastures	Grazing areas	Multipurpose woody trees, fodder for grazing animals and animals are also present.
Plantation cops with pastures and animals	Areas having less pressure on plantation crop lands	Plantation crops, animals and fodder for grazing animals.
Agrosilvopastural systems		
Multipurpose woody hedgerows	In humid and subhumid areas	Shrubs and trees.
Aquaforestry	Lowlands	Trees and shrubs that are preferred by fish.
Apiculture with trees	It depends on the feasibility of agriculture	Honey producing components are present.
Multipurpose woodlots	Various ecological zones	Multipurpose species are present.
Home-gardens involving animals	All ecological regions	Fruit trees and woody species are present.

Areas Suitable for Agroforestry System

The areas that are suitable for agroforestry include:

- Field boundaries.
- Cultivable lands.
- Cultivable waste.
- Old fallows.
- Other areas where there are chances for the incorporation of agroforestry.

Applications of Agroforestry

Resource Conservation

Conservation of natural resources like water, plants and lands can be carried out by watershed management. Conservation is an important aspect since it meets the socio-economic needs of human beings. However, degraded land can be conserved or improved by agroforestry practice. It reduces run off, controls erosion, enhances water table, and improves profitability, productivity and soil-moisture conservation.

Farming System

Natural resource management can be carried out via agroforestry practice. Therefore, with the change in economy, there is a need to address those technologies that are related to the integrated production of crops, livestock, fisheries and perennials. Diversification can lead to major shift in agricultural production. The shift in agricultural production can be boosted due to urbanization, budding of personal incomes and incorporation of world markets.

Biodiversity Conservation

Innovation in agroforestry contributes towards biodiversity conservation via conservation-development approach.

Carbon Sequestration

Agroforestry system serves as mitigation strategy and can also be employed for carbon equestration. Agroforestry provides a cost-effective means through various land use practices for carbon sequestration.

Enhancement in Soil Fertility and Structure

Plantation of woody species results in the enhancement and improvement of soil fertility. The reasons for an increase in soil fertility is:

- Leaf litter leads to the addition of organic matter of soil.
- Effective utilization of nutrients and efficient nutrient cycling.
- Nitrogen fixation by micro-organisms.

Criteria for Designing Agroforestry

The following criteria should be fulfilled for a good agroforestry system:

- Sustainability: With an improvement in sustainability of agricultural production system, the conservation goals can be achieved by agroforestry.
- Productivity: Productivity with agroforestry can be improved by various ways like:
    - Improvement in the yield of associated crops.
    - Enhancing the labour efficiency.
    - Diversifying the crop production.
- Adoptability: Technical design of agroforestry does not matter. The only thing that matters is adoption of that design by the farmers. It clearly means that this technology should fulfil the environmental and social characteristics of the land use system.

Energy Forestry

Energy forestry is a plantation of fast growing trees with the aim of producing wood as a renewable raw material within short rotation times, in which a fast-growing species of tree or woody shrub is grown specifically to provide biomass or biofuel for heating or power generation. If this happens exclusively for the production of energy, it is also referred to as energy wood plantation or energy forest.

The two forms of energy forestry are:

- Short rotation coppice.
- Short rotation forestry.

Short rotation coppice may include tree crops of poplar, willow or eucalyptus, grown for two to five years before harvest. Short rotation forestry is crops of alder, ash, birch, eucalyptus, poplar, and sycamore, grown for eight to 20 years before harvest.

Density Effect

These coppices contain about 1,000 to 4,000 stems per hectare or even 10,000 to 20,000 plants per hectare for dense coppice with very short rotation. A balance is to be found by the agrosilverist between density and desired stem diameter. In general, the biomass harvested per foot and the diameter of the stems (basal area) decrease with the density of the seedling. The big stems favor the woodcut and the fines interest rather the grinding. "A density close to 2,000 cuttings per hectare is currently the ideal density to optimize the production of ridges for the crushing industry".

Tree Species

In a short rotation plantation fast-growing and ausschlagfähige tree species, especially poplars (genus Populus) or pastures (genus Salix) are used. Research and breeding focuses on tree species that are suitable for temperate climates.

Among the poplars are especially the Balsampappeln such as the Western Balsam poplar and Populus maximowiczii and their hybrids and crosses between balsam and black poplar trees suitable. These are distinguished from pure black poplar and other tree species in that they grow very fast, do not need too much light and form a dense root system. Other targets in poplar combination breeding include increasing biomass production, improving pest resistance, accepting higher planting density, high cane rash, and rapid growth in the early years of development to reduce harvest intervals. The fuel quality of the wood is also influenced by breeding by selecting forms with high efficiency of nutrient use and high, but low-nutrient rootstock.

History and Economic Situation

A historic forerunner of the short rotation plantation in the coppice, in the trees in regular intervals u. a. for the production of firewood was put on the stick.

The cultivation of fast-growing tree species for material use for the production of wood-based materials has been successfully practiced in many European countries for many years. In Germany, however, wood fields have not reached beyond the trial cultivation stage since the 1970s, although the first research center for fast-growing tree species was founded in Hessen in 1974. On the one hand, this is due to a hitherto limited market for the end product wood chipsdue. On the other hand, heating oil has been the most economical fuel for heat supply for years. However, this situation changes as prices for fossil fuels increase. Today, short rotation plantations are mainly grown for energy use.

Since about 1990 willow clones grown in large plantations. In Sweden there are many years of experience with the cultivation of pastures in short rotation. As early as 1999, 16,000 hectares were cultivated there, in the 2005/2006 marketing year there were about 15,000 hectares of woodland in cultivation. In the newer varieties, the yield could be massively increased compared to the first wood fields and currently stands at an annual increase of 8 to 12 tonnes dry matter (atro) per hectare. In Germany, according to IACS, 5,968 ha of short rotation plantations were cultivated in 2014. Short rotation plantations are a very labor intensive cultivation method in the field of agricultural production. Most of the work can be done by farmers or foresters with their own machines. Only for planting and harvesting special machines are needed, which are usually provided by service providers.

Economic Vocation

The economic vocation, possibly alternative to agricultural activity, 19 is the most cited issue of energy crops; the TCR can contribute to producing fuelwood (heat or electricity production 7), poles, woodchips, fibers and to create or maintain rural or peri-urban jobs (through maintenance, exploitation, and marginally for monitoring and evaluation, qualitative and quantitative).

The sustainability of the TTCR sector is discussed (like all crops, the intensification of exploitation may lead to the depletion or overexploitation of the substrate by the export of nutrients 20 or the water resource).

It seems possible to create or maintain a renewable resource available for the wood-cellulose sector, or heat networks, with indirect spin-offs from tourism, and the exploitation of all or part of this afforestation.

Related Interest

In addition to their interest in energy crops, these wooded elements could – with appropriate management and if they are inserted into the landscape with a concern for ecological coherence – develop other interests for water, soil, air, water landscape, agriculture, health, carbon sink restoration, green and blue weave, etc.

For Breeding

The TCR can produce a part of the benefits of the grove (shading, protection against the wind, microclimate) for the breeding of mammals but also of poultry (Label chickens in particular 21).

Possible Landscape Vocation

If the species used are local species, adapted, varied, harmoniously planted according to the ecopaysager context (and thus hydrological and geopedological), by marrying the forms of the landscapes, and why not in the context of a neobocage semi-agricultural, hiding possibly black spots in the landscape (infrastructures and unsightly buildings), and if they are not exploited according to the principle of clearcutting but for example in several bands being rotated differently, so as to always maintain the equivalent of a hedge in the landscape, they can contribute to the restoration efforts or improvement of certain landscapes (A test with a triple cut woodleaf, each band being cut to a different year) was conducted in Belgium with a hedge of 100 m made up of 6 rows of willows, with an energy vocation of the harvest).

By reintroducing possibly elements of more natural appearance in peri-urban areas, or even in industrial zones, they can improve the landscape and connect to a green flow, a wild or "ecological" garden, a biological corridor, a green wall, a vegetated terrace, a landscaped park, etc.).

A secondary benefit in terms of cynegetic and rural tourism could then be expected, thanks to the improvement in terms of landscape.

Vocation to Protect Water and Fight Against Soil Erosion

Some effectiveness is recognized in afforestation and wooded strips in terms of water purification (for phosphates and nitrates in particular, but also in terms of reduction of turbidity). TCRs have been experimented in the tertiary phase of natural lagooning (for example at Lallaing and the wastewater treatment plant (for example in Villeneuve-d'Ascq for a few years), and fast-growing trees evapotranspired a large quantity of water. must be taken into account for their eco-landscape integration.

Usage

In the short rotation plantation so-called cuttings, Sections of annual, well-developed shoots, planted in rows and harvested mechanically or motormanually every 3 to 10 years, depending on species and climate. The rootstock remaining in the soil has the ability to cattle rash, and so after the harvest is a renewed sprouting of the trees, without having to be replanted. Under good conditions, short rotation plantations can produce on average about 10 tons of dry matter per year per hectare, and twice as much under very good conditions. After about 20 years, the production capacity of the plants decreases, and the area should be recreated with further use.

Cultivation

When planting short rotation plantations, deep soil tillage is strongly recommended by min. 25 cm deep plowing. This allows fast root growth and high wood growth rates. A good establishment success with growth heights of more than 1.5 m in the first growing season usually only sets in when, in addition to the conscientious soil preparation, the culture in the first growing season is kept largely free of weeds. This can usually be ensured by the use of pre-emergence herbicides directly after planting, as well as at least one mechanical care passage. A good culture care in the first year of growth improves the competitive situation of energy crops as well as the survival rate and the yield potential significantly. At the end of the second growing season, the crop closes already at most sites if the crop was kept free of weeds in the first two years.

As a rule, short rotation plantations are created in spring (end of March to beginning of June). As planting material are either pre-produced 20 cm – 35 cm long cuttings, which are planted by hand or by hand, or about 2.50 meter long rods, which are cut into about 20 centimeters long cuttings during planting in special planters. When planting it is important to ensure that the cuttings are placed as vertically as possible with good bottom closure in the soil. As a rule of thumb, the more difficult the location, the longer the cuttings should be. Short cuttings are planted at ground level, longer cuttings (from 30 cm) can look to about 1/3 from the ground. At very difficult locations it is recommended to plant plant rods> 100 cm in length, as these are much more robust to weeds and easier to care for. For cost reasons, the cultivation of plant rods is recommended only for a longer rotation in quantities of 2,000 – 3. 500 per hectare. Plant rods are planted at least 50 cm deep. The cultivation of short rotation seedlings is not carried out because the seeds of poplars and willows are not storable and are often no longer germinable after a few weeks. A much more expensive alternative to the cuttings derived from high yielding mother plants is the use of tissue culture propagation. However, this is currently not profitable and is used accordingly only in research. Short rotation plantations are usually created with special planting machines. All leading suppliers of planting material also offer the planting performance.

The planting association depends on the tree species, the planned product (energy or industrial wood) and the cultivation of short rotation seedlings is not carried out because the seeds of poplars and willows are not storable and are often no longer germinable after a few weeks. A much more expensive alternative to the cuttings derived from high yielding mother plants is the use of tissue culture propagation. However, this is currently not profitable and is used accordingly only in research. Short rotation plantations are usually created with special planting machines.

Plant Protection

In short rotation plantations, the same problems as in forestry crops occur. In addition to general dangers, specialized pests play a certain role, especially in individual species.

The problem with the cultivation of short rotation plantations is the low genetic variability of the plants, which results from the vegetative propagation of the cuttings, as well as the crowding of the plants. This makes the entire population susceptible to pests, with so far only the infestation by rust fungi of the genus Melampsora and by poplar leaf beetle (Chrysomela populi) led to greater yield losses. Against insect pests effective insecticides from agriculture and forestry are available, the use of which, however, so far only in rare cases.

In addition to mechanical plant protection methods such as the fencing of cuttings against Wildverbiss herbicides from crops are used against weeds during the establishment phase. For pesticide applications on unindicated crops, a derogation from the competent authorities is required in Germany under § 18b of the Plant Protection Act. This is usually given easily, as woodchips do not get into the food chain.

Under certain conditions, it is necessary to completely or partially dispense with the use of herbicides. For example, this is the case in water protection zones or in organic farming. There are several alternative measures that can be taken: early and multiple mechanical care and mulching films allow the operator to prevent too much accompanying vegetation. Mulch films reduce the care measures and offer the trees protection and clear growth advantages. Pastures as fast-launches are particularly suitable, since unlike poplars they tend to do without care measures. Mice help against mulching between the rows and birds of prey on the surface.

Harvest

Short rotation plantations are permanent crops that can usually be used for up to 20 years with an average of 3 to 6 harvest rotations. The harvest is carried out over several years Umtrieb, the stock has at harvest time reaches a height of 6 to 8 meters. The yields for balsam poplars are between 10 and 15 tons per hectare per year, for pastures 5 to 10 tons per hectare per year. The harvest time is optimally in the winter after the leaf waste, as this can leave a large part of the nutrients in the fields. As harvesters are mostly high-powered, self-propelled with short rotation Forage used with small timber feeder. Harvester and bundlers from the forestry technology can also be used in higher-yielding crops as well as longer rotation times. It is important not to over-compact the soil so that the trees can stand in a well-ventilated soil.

Fresh wood chips usually have a water content of 50 to 60% and must be pre-dried for use. This happens when the chips are stored outdoors or in pressed bales. For firing in smaller and medium sized plants they should have a maximum water content of 30%. Important is a rough hacking of the wood, as fine wood chips are poorly storable. The chips can be stored as rents and burned, gasified or processed into pellets as needed. Short rotation plantations drive off again after the harvest. Depending on the nutrient supply of the soil, a fertilizer of 50 to 70 kg nitrogen may be used after the harvest/ ha makes sense, but usually no fertilization is necessary because hardly any nutrients are removed from the surface via the wood.

Profitability

Compared to established one-year arable crops, KUP require relatively high investment in founding existing crops and a relatively long production period and generate irregular cash flows. At the same time, after a successful establishment, there are no significant additional costs for the management of the plantations, apart from the harvest costs. Finally, at the end of the useful life, the cost of reconversion of the area for arable use (removal of rhizomes) must also be calculated. CUPs therefore represent a greater entrepreneurial risk than one-year crops; cultivation seems to make economic sense only for fixed purchase contracts with binding prices.

In order to compare the economic efficiency of KUP with that of one-year crops, it is advisable to use the dynamic investment calculation instead of the contribution margin ; The irregular cash flows can thus be converted into an annual constant annuity. The calculated annuity can be compared directly with the contribution margin of one-year crops. According to various studies, the annuity of poplars and willows may, under favorable conditions, be around € 250 to € 500 per hectare per year, but in unfavorable conditions also be negative. The higher the number of rotations, with a rotation period of three to four years, the sooner positive annuities are achieved.

A profitability analysis of the cultivation of fast-growing tree species in short rotation plantations based on current literature and practice data in 2014, shows that it is economically profitable in the majority of cases, and from an average yield level of 11 to 12 tons of absolute dry matter per hectare per year can certainly compete with agricultural crop rotations. Compared with the average yields and prices of a rotation typical for middle axles (winter rape – winter wheat – winter wheat – winter barley) from 2006 to 2010, a farmer with energy wood production generates an average profit of 45 per ha. Additional economic benefits can be provided by activating short rotation plantations as "ecological priority areas".

The total cost of establishing a short rotation plantation is around 2,000 to 3,000 euros per hectare. The first harvest of poplar and willow plantations usually yields significantly lower yields than in subsequent years. From the 2nd harvest, the annual growth rates are i. d. R. at least 50% higher compared to the 1st harvest. In order to keep the specific harvest costs low, it is advisable to wait with the first harvest of a short rotation plantation until a total harvest of at least 25-30 tons of dry matter per hectare is expected. However, care must be taken to ensure that the trunk thickness of the trees does not exceed the possibilities of the harvesting technology used.

In the heating market, woodchip firing becomes economical due to the sharp rise in fossil fuel prices, even at higher investment costs for the combustion and charging technology. The average energy content is between 15.5 and 18.5 MJ / kg, while harvesting and comminution consumes about 0.06 MJ / kg, which makes the energy balance very good. It should be noted, however, that fast-growing plants also have lower specific energy content, so that the volume of raw materials to be processed increases sharply. For example, a room meter of oak firewood supplies approx. 1890 kWh, fast-growing poplar only 1110 kWh. This is not clear in the calorific value, since it refers to the mass and not to the volume.

Short Rotation Plantations Devastated and Urban Areas

The global increase in food prices in 2007 and 2008 has led to a discussion on whether and to what extent the production of renewable raw materials for energy production has fueled this increase. For this reason too, pilot projects have been started for the plantation of short rotation plantations on land that is not in competition with food production. In this context, experiments in devastated areas of former opencast mines and on inner-city areas deserve special mention. The latter were started in 2007/2008 by Stadtwirtschaft Halle on areas that were previously occupied by residential buildings that were demolished as part of the so-called "Urban Redevelopment East" in Halle (Saale).

Climate Protection Effect

Bioenergy value chains based on short rotation plantations for electricity, heat and fuel production enable a significantly higher CO2 avoidance compared to traditional bioenergy processes. This is due to the extensive management, which means that the input of energy in the form of fertilizers and pesticides or machine use is extremely low. Also the CO_2- Avoidance costs are only a fraction, compared to biofuels from oilseed rape and grain or biogas from maize silage. Use of biomass for energy – recommendations to the policy: "The policy could by a change of course in the promotion reach (more biogas based on gül and electricity and heat based on wood chips KUP) that, with consistent use of resources and land, the contribution of bioenergy to climate protection would more than triple. "

In addition, short rotation plantations offer the opportunity to manage arable land, which has a particularly high drainage requirement. When using willow and poplar clones, therefore, an excessive drainage of these surfaces can be dispensed with, which has a clearly positive effect on their CO2 balance? If particularly humid locations are used in this context, a possible negative influence on the local water balance and thus on sensitive ecosystems can be avoided.

Influence on Biodiversity

When assessing the impact of short rotation plantations on biodiversity, it is crucial to decide which reference system is used. Most of the studies come to the conclusion that short rotation over intensive agricultural use has a positive effect on (plant) biodiversity, but has low biodiversity compared to near-natural forest. In this assessment it should be noted that the conversion of forest into short rotation plantations is strictly prohibited and cultivation is concentrated exclusively on arable agricultural land. The NABU notes that the system of SRC in agricultural regions contributes to the structuring of the landscape and the creation of habitats and stepping stones for plant species. For a comparison with the Grassland use was still lacking robust investigations. For biodiversity in terms of animal species, on the other hand, CUPs would not be of much importance, since endangered species are scarcely available in intensively used regions. However, SRCs would benefit animal ecology, as animals benefited from a longer period of dormancy, less fertilizer and pesticide use and fewer disturbances compared to one-year crops.

Benefits

The main advantage of using "grown fuels", as opposed to fossil fuels such as coal, natural gas and oil, is that while they are growing they absorb the near-equivalent in carbon dioxide (an important greenhouse gas) to that which is later released in their burning. In comparison, burning fossil fuels increases atmospheric carbon unsustainably, by using carbon that was added to the Earth's carbon sink millions of years ago. This is a prime contributor to climate change.

Compared to other energy crops, wood is among the most efficient sources of bioenergy in terms of quantity of energy released by unit of carbon emitted. Other advantages of generating energy from trees, as opposed to agricultural crops, are that trees do not have to be harvested each year, the harvest can be delayed when market prices are down, and the products can fulfil a variety of end-uses.

Yields of some varieties can be as high as 12 oven dry tonnes every year. However, commercial experience on plantations in Scandinavia have shown lower yield rates.

These crops can also be used in bank stabilisation and phytoremediation. In fact, experiments in Sweden with willow plantations have proved to have many beneficial effects on the soil and water quality when compared to conventional agricultural crops (such as cereal).

Ecological Impacts

They are still partly discussed. In 2010, there are still few advanced studies on the ecological effects of this coppice on biodiversity (45 references found by Gosselin in 2009). Some TCR and TTCR plantations have been made for research and demonstration purposes since the 1980s, but apart from Gustafsson's (pioneering) isolated work, it was not until the 1990s (in the United States) and 1995 in Europe to find scientific publications, particularly in Sweden and the United Kingdom, the decline is older.

Very short, short or medium rotations are currently generally high-density plantations (10,000 to 15,000 stems / ha for TTCR and 1,000 to 4,000 stems / ha for TCR), monospecific in variety. hybrids (more productive, but often more fragile), even monoclonal or genetically little biodiversified or even exotic.

While they can – under certain conditions – help purify water and soils, willow and / or poplar coppices are very water-intensive in Northern Europe (up to 6-7 mm / day), in summer when this resource is the smallest.

A permanent coppice (but can be partly periodically cut) can provide protection to a part of the biodiversity or contribute to ecosystem restoration, but the available studies show that they benefit rather mundane species.

Their uniform appearance (same age class) and the absence of wood-dead stage diminish their ecological interest and sometimes make them ecological pitfalls. They attract birds, insects and mammals that take advantage of them for some time and then die when they are crushed with wood or brutally left without habitat. In addition, in the long run, these crops may require large amounts of inputs and exploitation makes them susceptible to various diseases.

However, a more heterogeneous structure (in age classes) and more diversified (in species and genes, although mixing clones induces in the " dominated clones " phenomena of mortality by competition, especially when the plantation is very dense could perhaps reduce the need for chemical and phytosanitary inputs by making the practice of energy culture more flexible (less input costs, less maintenance, better resistance to storms, frost, droughts); a culture of different species, densely planted, would develop roots colonizing different depths and make the crop more resilient and resilient.

Finally, according to the available data, the ecological effects of this coppice vary greatly according to the eco-landscape context and its dominant (agricultural, forest, urban) and according to the former state of the soil and the environment (former annual agricultural crops, fallow land, grassland, wastelands, possibly polluted, forest).

Post-Ecological Aspects

If they are not subject to phytosanitary treatments, if they do not substitute for natural environments that are richer and more interesting for biodiversity, and if they have genetically and structurally complex multistrate coppices connected to the local green network could be of some ecological interest.

Remediation of natural environments or natural infrastructures to fight against the fragmentation of natural habitats which we know has become one of the first causes (and probably even the first cause) of regression and disappearance of species, for example if one thinks of integrating ponds or ditches in water, and if one avoids the phytosanitary treatments. Multistrate linear plantations can – to some extent – play an ecological infrastructure role (= biological corridors) which are semi – natural, but which may play a role as an alternative biological corridor.

Regulation and storage of water resources (very favorable impact on flood / drought cycles by inertia and regular supply of groundwater by natural drainage). Microclimatic regulation (macro and micro) and thermohygrometric regulation in particular. Absorption of greenhouse gases, carbon sink (in soil), to some extent Favorable impact on the maintenance of natural balances (for example by increasing ecotones or creation of buffer zones and the increase of semi-natural habitats (if the coppice or crops are not subject to pesticides and excesses) fertilizer possible (locally) reduction of the overcrowding of natural environments (often relictual).

Detoxification of some polluted wastelands (e.g. by nitrates / phosphates or even by some heavy metals), but with a necessary prudence, the toxic ones possibly reappearing in the foliage, the woods, the fumes or the ashes. The wood may need to be burned in installations with efficient filters. Absorption or fixation of certain atmospheric pollution (dust, gas, aerosols).

Adverse Ecological Aspects

They appear especially when the coppice replaces an environment of higher ecological value (permanent meadow, peat bog, heath, limestone hill, floral fallow, etc.), but in general the TCR and TTCR attracts rather banal species, some of which go die crushed or crushed at harvest.

The species used (Salix sp., Populus sp.) In TTCR are among the most water consuming (up to 6-7 mm / day) and often requires (to be industrially exploitable) tillage, control of competing plants (usually by means of a herbicide) and a stump to cause the floor to the other uses. Pollen, seeds and clones releases can be used to source genetic pollution for wild and native populations near.

Pedagogical Vocation

The TCR can be a place for training, awareness, formation of communities, elected officials, technicians, associations, schools and public, a place of demonstration of new methods of reconstitution, maintenance and more "ecological" management of Para-forest or semi-natural ecosystems, a place of "Chantier – école"? Alternate training plans are relatively easy to implement (given the seasonal nature of the work).

Scientific Vocation

The TCR can support the development of research on a new form of ecological engineering (reconstitution of semi-natural or semi-forest ecosystems, soil release of ashes or certain trace elements, sulfur, for example; evolution of these semi-natural environments and the ecosystems that appear in them, evaluation of the different methods, their relevance over time, or depending on the pedological and climatic contexts, ecotoxicology (effects of the TCR on soil purification, migration of toxic, etc.); reintroductionor in situ conservation of "useful" plant and animal species; design of reproducible methodologies and techniques; monitoring of mycorrhizal phenomena in this particular context, etc.

Problems

Although in many areas of the world government funding is still required to support large scale development of energy forestry as an industry, it is seen as a valuable component of the renewable energy network and will be increasingly important in the future.

Growing Trees is Relatively Water Intensive

The system of energy forestry has faced criticism over food vs. fuel, whereby it has become financially profitable to replace food crops with energy crops. It has to be noted, however, that such energy forests do not necessarily compete with food crops for highly productive land as they can be grown on slopes, marginal, or degraded land as well – sometimes even with long-term restoration purposes in mind.

Sustainable Forestry

Sustainable forestry is about managing forests to provide wood, clean water, wildlife habitat and beautiful landscapes.

Fig: Tree seedling in forest.

"Sustainable" means to maintain, continue, and keep, while "forestry" is the science and art of managing forests. Thus, sustainable forestry is about caring for and managing forests to provide the natural resources, such as wood and clean water, we need now and in the future. It also means sustaining other things we value from the forest like wildlife habitat and beautiful landscapes. Sustainable forestry is concerned with all parts of the forest--trees, smaller plants, soils, wildlife, and water. It involves protecting forests from wildfire, pests, and diseases, and preserving forests that are unique or special.

Fig: Protecting the quality of water in the forest is critical to all aquatic and terrestrial life.

Sustainable forestry also puts an emphasis on people. People need forests for the resources they provide, and as a place to live or to make a living. Sustainable forestry ensures that future generations will have forests to meet their needs and values. As you can see, sustainable forestry is complex and can involve many things. Let's look at a few of these things more closely. You will also see how sustainable forestry can mean different things to different people.

Fig: Sustainable forestry involves forward thinking. It's just as important to think about which trees will be left after a harvest as which trees will be cut.

Forest Management

To care for and manage a forest in a sustainable way it is necessary to use responsible management practices. These are often specifically adapted to each site. One of the most important practices is to look at whether the forest has enough natural seeds, seedlings, and tree sprouts (all called regeneration) to make a future forest. Excessive populations of white-tailed deer in a forest greatly reduce regeneration by eating young trees. Too many ferns or too little sunlight can also play a role. Many sustainable forestry practices can protect or encourage forest regeneration. They include putting up a fence to exclude deer, controlling weeds and other plants, and removing some trees to allow more sunlight to reach down into the forest.

When it comes to removing trees, or timber, from the forest, many practices can assure the forest's future. It is a good practice to think as much about which trees to leave as about which trees to cut. The trees left in the forest will continue to grow, occupying and regenerating the forest for many years to come. If they are chosen properly, the remaining trees can provide many of the same values and resources, and, perhaps, new ones in the future.

Other sustainable forestry practices include protecting forest streams and wet areas. Harvesting trees can disturb and expose soil in small areas. This is especially true on roads built for driving the machinery used to remove trees. The flow of rain or other water across, and under, roads must be carefully managed with culverts (large pipes) and proper road design. This helps prevent soil from washing into a stream or wet area. Too much soil in water, or sediment, is harmful to aquatic life. Trees and other vegetation left undisturbed adjacent to streams or wet areas can also prevent soil from entering streams. These areas are called buffer strips.

Browsing Concerns

Forest Health

Sustainable forestry is also about maintaining forest health. Many living and nonliving things make a forest unhealthy. These include insect pests, diseases, wildfires, and extreme weather. Normally, the loss of individual or small groups of trees is not a big concern. When larger numbers of trees die, it may signal a forest health problem. Sustainable forestry seeks to promote forest health by reducing things which seriously damage the forest.

Fig: Assuring that a forest has enough new trees to regenerate after a harvest is an important practice. Deer exclusion fencing (top) and tree tubes (bottom) can protect seedlings and stump sprouts from deer browsing.

Fig: Gypsy Moths are just one of the pests that can impact forest health. They must be monitored and sometimes controlled to protect forests.

Caring for Special Sites

All forests are special, but some areas are more unique than others. It is a principle of sustainable forestry to protect unique areas that are not found elsewhere or are uncommon. One example is very old forests. Almost all of the 17 million acres of forests were clear-cut at least once by the early 1900s. However, a few small areas were never cut. Most of these old forests are currently preserved.

Fig: Pennsylvania has very few old growth forests left. Most of these are protected for future generations to enjoy and study. Some are over 500 years old!

Different Approaches

People view sustainable forestry in many different ways. To a forest landowner, sustainable forestry might mean selling timber and yet passing the land to an heir in good condition. To a logger (a person who cuts trees for a living) sustainable forestry might mean protecting the trees left in a forest, and constructing roads properly. To a professional forester, sustainable forestry might mean selecting appropriate forestry practices for a given site. All of these people want to sustain forests, even though each views the forest in different ways. Sustainable forestry is a broad but important concept. It requires concern and commitment on everyone's part.

Ecoforestry

Ecoforestry is about forests -- about protecting the forest and related ecosystems that sustain life on Earth. Ecoforestry is not about canny ways to cut more trees or clever ways to combine logging with berry-picking and bird watching. Ecoforestry is about understanding our place as humans in ecosystems and acting on that understanding. Ecoforestry is about future generations of people -- about taking the time to provide intergenerational equity for all life.

Ecoforestry is ecosystem-based. The goal of ecoforestry is to maintain fully functioning forest ecosystems. In contrast, the central goals of conventional forestry are to maintain timber supplies and an adequate return on stockholder*s investments. Conventional forestry, in other words, is timber-based or profit-based. Human societies depend (more than we usually care to admit) on healthy, fully functioning ecosystems. Because ecoforestry maintains whole ecosystems, it also maintains healthy human communities. When forests are degraded, human communities that depend on the forest suffer a decline in every measure of well-being.

Human economies depend (more than we usually care to admit) on healthy, fully functioning human communities. Because ecoforestry maintains human communities, it provides the options and foundations for a sustainable, diverse economy, based upon meaningful work, which provides for the whole person -- spiritually, mentally, and physically.

Conventional forestry provides jobs; ecoforestry provides jobs that people love to do. While cutting fewer trees, ecoforestry can, in fact, provide more jobs than conventional forestry (by a factor of three times or more) by creating not only a wide spectrum of timber and wood product-related jobs, but an even wider spectrum of non-timber jobs, from wildcrafting to permaculture to ecosystem restoration. Because ecoforestry provides for a continuation of large, old trees, ecoforesters will always be able to supply high quality wood that is necessary for value-added wood products, from timber frames to furniture. Such diverse, community-based economies are the only sustainable economies in the long term.

Albert Einstein provided some wisdom that is applicable to ecoforestry: "A clever person solves a problem -- a wise person avoids it." The lifetime of a tree starts with the germination of a seed and ends when the fallen tree finally disappears into the forest soil. This lifetime may span 500 to 2,000 years and beyond. Forests are built upon fallen trees, yet conventional forestry does not replace them. Without waiting centuries or millennia for natural processes to occur, no person is clever enough to replace fallen trees. However, ecoforestry gives us the wisdom to avoid this and other problems.

Principles of Ecoforestry

- Sustainability is the most important factor in Ecoforestry. Therefore it must be the first consideration in any planned operation. By doing this, one ensures the protection of such things as rare species, sites of native cultural significance, riparian zones, etc.

Fig: Ecoforestry Ecoplan.

- There should be no impact on Riparian zones. This protects the water quality by not altering the drainage pattern of the zones. Therefore no tree removal should take place in the most sensitive areas.
- The composition and structure of a forest is vital. So encouraging natural regeneration and retaining senescent trees maintains the age and vertical diversity of the ecosystem.
- When removing trees, insist on low-impact harvesting methods. As a result this minimises disturbance and avoids soil compaction.

Fig: Ecoforestry Ecoplan Forestry.

- Prohibit clear-felling. Clear-felling is not ecologically correct, and in many cases is not financially the best option anyway. Adhere to continuous cover management systems to ensure biodiversity benefits and long-term sustainability.
- Carefully select trees as candidates for removal by considering your long-term goals. Hence retained stems should be chosen based on their native provenance, habitat value, seed

production or even aesthetics. Consider how these trees will benefit the ecosystem into the future, and plan around that.

- Allow forests to regenerate naturally from existing seed-trees. This removes the need for expensive planting and allows the forest to develop naturally with species that are ideally suited to the site and local conditions.
- Limited or preferably no pesticide use. The forest requires disease, insects, and competition, as a result they are essential parts of a fully functioning forest.
- Retain deadwood and fallen branches wherever possible, because these provide additional habitat and restore soil quality over time.

Fig: Ecoforestry Ecoplan Forestry Ltd.

- Maintain the aesthetics, beauty and unique qualities of the woodland.
- Always look at the forest as a whole, from the soil and humus underfoot to the highest canopy.
- Rely more on people and niche markets. Use alternate thinking to depend less on the destruction or harvesting from the forests.
- Finally, be enlightened. Don't allow ignorance to limit what you can do, recognise that the ecosystem needs sensitive management. If the forest is not preserved, then it cannot be harvested forever.

Wildlife Forestry

Wildlife forestry is management of forest resources, within sites and across landscapes, to provide sustainable, desirable habitat conditions for all forest-dependent (silvicolous) fauna while concurrently yielding economically viable, quality timber products. In practice, however, management decisions associated with wildlife forestry often reflect a desire to provide suitable habitat for rare species, species with declining populations, and exploitable (i.e., game) species. Collectively, these species are deemed priority species and they are assumed to benefit from habitat conditions that result from prescribed silvicultural management actions.

Early wildlife conservation efforts largely focused on controlling indiscriminate slaughter of wildlife by restricting the season or sex of harvested species. Subsequent conservation efforts targeted increased protection of populations through creation of sanctuaries or reserves, such as national parks and wildlife refuges, within which harvest of wildlife was prohibited or greatly restricted. Typically parks and wildlife reserves were located in areas with abundant wildlife populations, and therefore, little emphasis was placed on active management to improve or maintain suitable habitat.

In his description of a "land ethic" in which people are members of the natural community, Leopold indicated the appropriateness of habitat alteration and management to aid the continued existence of wildlife species. His philosophical principles were succinctly stated as "A thing is right when it tends to preserve the integrity, stability, and beauty of the biotic community. It is wrong when it tends otherwise."

Anthropogenic alteration of forest habitat for the benefit wildlife species has long been practiced, particularly with regard to intentional use of fire. Even so, active manipulation of forest structure to purposefully enhance wildlife habitat has been slow to evolve. Most early efforts regarding habitat alteration focused on providing suitable escape cover, foraging habitat, or distinct habitat features (e.g., nest sites) for game species. Within forested landscapes, improving wildlife habitat often meant increasing local heterogeneity and providing more forest edge habitat. However, increased knowledge of the possible negative effects associated with forest edges, such as greater predation and nest parasitism rates, has called into question the practice of providing increased heterogeneity within forested landscapes. At the same time, a desire to provide habitat for specific species that may require specialized habitat conditions, as is the case for some endangered and threatened species as well as other charismatic megafauna with widespread public support, has prompted development of habitat management prescriptions designed to improve forest habitat conditions for specific species.

Partly in response to what has been perceived as management for single-species, but also to advocate a desire to prevent future population declines of common species, recent emphases in wildlife conservation have been placed on the development of comprehensive, eco-regional, conservation plans whose scope encompasses multiple species or the entirety of a species-group. This approach gained momentum with the North American Waterfowl Management Plan (1986) which identified conservation actions intended to return waterfowl populations to their former abundance and distribution. Subsequently, this basic concept has been expanded to other species groups such as songbirds, reptiles and amphibians, and shorebirds, as well as other species groups. Although these conservation plans differ markedly with regard to their recommendations for habitat management, many of these conservation plans advocate use of prophylactic management prescriptions to enhance or maintain suitable habitat conditions for priority wildlife species. Indeed, most conservation plans recognize the need for alteration of habitat and landscapes via management actions so as to attain the desired distribution, abundance, and viability of priority wildlife species.

Before managers can provide appropriate forest habitat for priority wildlife, the landscape and site characteristics that contribute to viable populations of these species must be identified. Preferably, those forest habitat characteristics identified can be quantified via standard measurement protocols. If possible, the desired or acceptable range of values should be determined, and either quantitatively or qualitatively stated, for each of these characteristics. In addition, a threshold value for each characteristic should be determine, that if exceeded would justify prescription of management actions to return site characteristics within the desired range.

Because wildlife forestry silviculture does not target optimal production of forest products, many forest management practitioners (i.e., foresters) have been reluctant to adopt these alternative silvicultural practices. Indeed, these concerns may be justified, as desired stand conditions may include less then fully stocked stand densities, retention of some economically mature trees, maintaining species diversity which includes less merchantable tree species, and preservation of less vigorous (decadent) trees. However, at present we lack credible, long-term economic data to suggest that development and maintenance of quality forest habitat for priority wildlife species will result in a sacrifice in timber production or quality. Regardless of these overarching financial concerns, silvicultural management actions prescribed to enhance desirable forest habitat conditions for priority wildlife species are expected to be commercially viable. Maintaining the long-term commercial viability of prescribed silviculture is import, because without sufficient financial returns, needed management actions will not be undertaken.

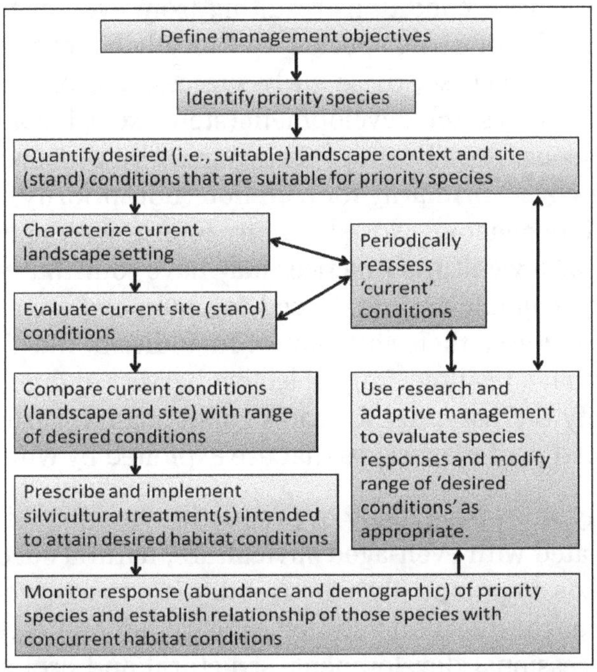

Fig: Characterization of wildlife forestry silvicultural methodology.

Forest Management

Silviculture is the manipulation of the establishment, growth, composition, and quality of forest vegetation to achieve management objectives. Historically a primary objective on managed forest lands has been to maximize financial returns from production of forest products such as timber and wood pulp. Production-oriented silviculture has focused on reducing the length of production cycles between stand regeneration and harvest while concurrently maximizing wood product yields. Reduction of production cycles (i.e., rotation length) tends to lessen the time regenerating forests are in early seral habitats, and consequently the size and age of shrubs and there their likelihood of persistence within subsequent seral stages is reduced. Often production-forestry objectives favored a limited number of tree species due to their high economic value or growth potential.

Concomitantly, to increase growth and yield of desired trees, other species with lesser economic value, as well as congeneric competitors, were culled. These actions result in increased spatial homogeneity within managed stands. The apogee of this trend towards increased spatial regularity is expressed within planted plantation forests. Climate, soils, species, and other factors affect the optimal length of production cycles within production- oriented forests. Thus, rotation periods within plantations vary markedly from rotations of up to 150 years to as few as 10 years. Notably, recent developments and increased advocacy of short-rotation woody coppice systems for use in biomass production have further compressed the length of production cycles to <4 years.

Even in non-plantation forest management systems, historical forest management practices, especially those focused on timber or pulpwood production, have often resulted in relatively homogeneous mature forest conditions. This homogeneity has been exacerbated by environmental changes that limit natural disturbances within forest stands (e.g., fire suppression). When these homogenous forests are managed using even-aged silviculture, during which all or most trees are harvested and the entire stand regenerated, successional 'boom and bust' conditions are dispersed throughout landscapes. The forest conditions resulting from even-aged management (e.g., clear cuts) are temporarily favorable for some species, specifically benefiting species that exploit early-successional forest habitat. Even so, these early-successional habitats are transitory and are followed by an extended period of stand development (stem exclusion and understory reinitiation; during which the homogeneous, closed-canopy forest structure with sparse understory vegetation offers attraction (forage or cover) primarily for common (non-priority) forest wildlife species. Indeed, the abundance of homogeneous, closed canopy forests with sparse understory cover that have resulted from historical silvicultural practices may have contributed to why species for which these habitat conditions are suitable are considered common and are not priority species in need of management actions to increase their abundance. In addition, the altered (often depauperate) species compositions that have resulted from selective harvest and culling during past management practices may amplify the potential for catastrophic damage from outbreaks of insects or diseases or limit the seasonality of mast crops that are exploited by wildlife.

In contrast to the relatively short periodicity of production-oriented silviculture and the periodic boom-bust cycles associated with even-aged silviculture, natural successional processes within forests may require hundreds, or potentially a thousand, years to achieve complete replacement via small-scale disturbance or as the duration between catastrophic stand-replacement events. Owing to differences in their successional development, structural and ecological characteristics within production-oriented managed forests and in forests managed using even-aged silviculture are manifestly different from those same characteristics within natural forests. Ecological differences between managed and natural forest stands are exacerbated over time, as natural stands approach climax or 'old-growth' conditions while intensively managed forests are repeatedly regenerated.

To mitigate the detrimental effects of forest management on priority wildlife species, recommendations have been made to increase the length of production cycles, thereby allowing increased time for development of heterogeneous forest structure. Unfortunately, small increases in rotation lengths provide little opportunity for increased complexity of forest structure within these stands – and periodic, complete harvest and stand regeneration thwart retention of forest structure over time. Similarly, forest managers have argued that temporal and spatial dispersion of clear-cuts throughout landscapes will provide continual availability of wildlife habitat – albeit at different locations. Although early-successional habitat may be continuously afforded via such dispersion, few of the other benefits conferred by increased forest structural heterogeneity, such as emergent or senescent legacy trees, are realized.

Increased recognition of the structural and ecological deficiencies within production- oriented forests has spurred interest in and development of alternative silviculture. Alternative silviculture has been especially valuable for stands where management objectives are not solely financial but where stewardship objectives include preservation of regional biodiversity, enhancement of wildlife habitat, maintaining landscape aesthetics, or providing increased recreational opportunities.

Alternative silvicultural methods developed for use in many forest types have been based on identification and quantification of natural disturbance regimes, with subsequent implementation of silvicultural methods intended to emulate these disturbances. For many forest types, emulation of natural disturbance regimes includes small-scale disturbances implemented via single-tree and patch-cut harvests. If successfully implemented, emulation of natural disturbance regimes should promote natural stand development and succession. Unfortunately, natural stand development and concurrent successional changes may conflict with landowner objectives if succession results in changed species composition or reduced financial returns. Moreover, strict adherence to silvicultural regimes that mimic natural disturbance may not be necessary to maintain biodiversity or to provide desired wildlife habitat.

Another common alternative silviculture has been inspired by the desire to achieve forest structure reminiscent of old-growth forest. Although the structure of old-growth forest varies among forest types, common attributes generally include increased vertical heterogeneity via a multi-layered canopy, increased horizontal heterogeneity associated with canopy gaps and different regeneration cohorts, the presence of large trees that are often in older-age classes, abundant snags (standing dead trees) or senescent trees, and large diameter downed woody debris. Myriad ecological benefits, including increased biodiversity, have been ascribed to old-growth forests. As such, forest reserve lands have been designated (e.g., wilderness areas and parks) which afford unfettered, natural development of these characteristics. However, on lands with timber production as an objective, even if not the sole objective, implementation of production cycles that span hundreds of years is not likely to occur. In addition, merely increasing the length of production cycles may provide little enhancement of biodiversity if the forest structure favored by production-oriented silviculture is maintained. Furthermore, where forest reserves are small or located within inhospitable landscapes, full ecological benefits of old-growth forests may not be attainable or sustainable.

Fortunately, it is likely that many of the benefits conferred by old-growth forests, or at least specific species associated with old-growth forests, are positively related to the heterogeneous structure (vertical and horizontal) and other structural attributes found within these forests, not necessarily with the prolonged existence of these forests. As such, alternative silvicultural methods have been developed which retain or enhance structural heterogeneity within managed forests, thereby promoting structural attributes of old-growth forests and encouraging greater biodiversity. Most often, increased heterogeneity and other desired attributes have been achieved via retention of living or dead trees (or other forest elements). These retained structures have been referred to as legacy elements.

Based on the concept of retained legacy elements, various alternative silvicultural methods have been advocated to enhance structural complexity within managed forests. Some of these methods have being designated as green-tree retention harvests, variable-retention harvests, dispersed retention harvests, variable density thinning, and active intentional management. The legacy elements retained under these different silvicultural regimes may differ markedly. Retained elements may differ in quantity (e.g., % canopy cover or % basal area retained) but they may also differ in their composition, as different canopy species are preferentially harvested or specific regenerating species favored for retention. In addition, the quality of retained elements may differ via selection of retained trees with differing perceived fates – from robust, canopy-emergent trees through decadent, dead or dying trees. Moreover, retained elements may differ in their dispersion and the spatial juxtaposition of retention (e.g., clumped or linear).

Wildlife forestry silviculture exploits the versatility of legacy retention by combining specific retention and removal of canopy to achieved habitat conditions that are deemed suitable for priority wildlife species. The resultant forest stand may exhibit a mixture of tree species, size classes, and decadence within a heterogeneous distribution of aggregated retention and canopy gaps.

Landscape Considerations

Although specific silvicultural methods have most often been developed for and are implemented within forest management units (a.k.a. forest stands), a similar ecologically based approach has evolved with regard to maintaining or restoring landscapes surrounding managed forest stands. Intact forested landscapes are conducive to maintaining floristic and wildlife diversity, but also provide tangible social and economic benefits.

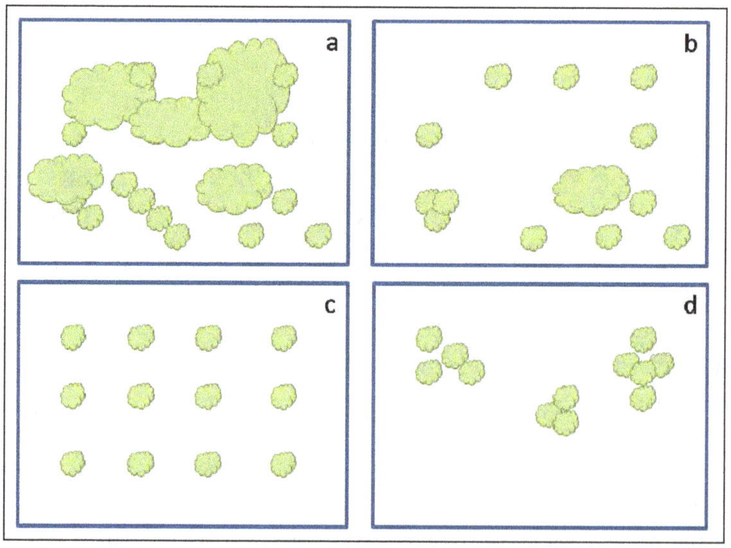

Fig: Potential distribution of legacy elements after application of alternative silviculture, differing in quantity of retained canopy cover (a, b) or dispersion of retained elements (c, d).

Loss of diversity within landscapes is often a reflection of the increased homogeneity within forest stands. Although forest structure within a landscape ranges from regenerating stands to mature merchantable stands, these stands may be monotypic in species composition and physical structure. In addition, ensuring appropriate size of habitat patches and maintaining connectivity among these patches is critical to sustaining viable wildlife populations within landscapes. Although the appropriate patch size is dependent upon the species under consideration, increased fragmentation tends to disrupt dispersion of fauna and flora and when habitat area is reduced without appropriate connectivity, extirpation of some species can result. Finally, maintaining landscapes capable of sustaining viable wildlife populations requires that evaluation, planning, and management decisions reflect the temporal processes inherent in the ecosystem.

Four landscape conditions have been recognized as contributing to ecosystems that support sustainable wildlife populations:

- Diversity within the landscape.
- Connectivity among landscape elements.
- Appropriate habitat patch sizes.
- Sufficient time to achieve ecological functions.

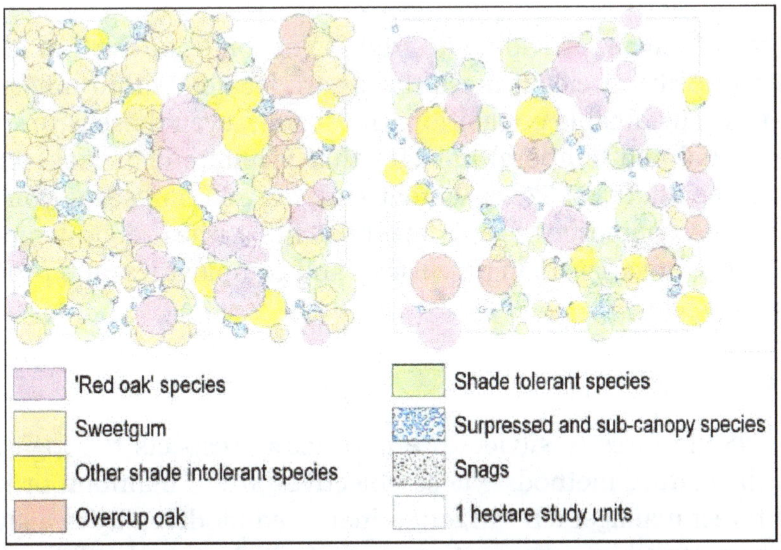

Fig: Canopy cover accounted for by individual tree crowns within bottomland hardwood forests on an untreated control stand (left) and on a stand subjected to wildlife forestry variable-retention silvicultural treatment (right), 6 years after treatment.

As when considering disturbance within the context of forest stand development, landscape conditions have often been influenced by large-scale disturbance events, particularly fires. Forest landscapes that result from management based on historical fire regimes, when compared with landscapes with extensive reserves and traditional silvicultural management had increased area of late-successional habitat, more over story structure in young forests, and larger forest patches. Similarly, when compared with human altered landscapes, natural landscapes were composed of larger habitat patches which formed the landscape matrix, had greater connectivity among patches, the range of variation in patch size and complexity was greater, as was the likelihood of encountering 'rare' habitat types.

Despite some understanding of desirable characteristics of forested landscapes, the appropriate landscape area remains nebulous. In part this is because an appropriate landscape area is dependent upon the species being considered, landscape characteristics, desired response, and the management context. As an example, the landscape appropriate for a grizzly bear is vastly different from the appropriate landscape for a shrew.

Even with knowledge of the appropriate landscape area and characteristics, manipulation of landscape conditions is usually beyond the purview of individual landowners. This is true even for large industrial or governmental landowners. Nevertheless, landscape conditions influence the ecological benefits conferred by forest stands, including the distribution, abundance, and viability of wildlife species within forest stands. In boreal forests, stand structure accounted for 54% of variability in bird and small mammal communities but landscape characteristics accounted for 29%. Similarly, bird species abundance within coniferous dominated landscapes was positively related to the proportion of broadleaf forest, with landscape thresholds for individual species ranging from 1% to 25% canopy cover. A strong negative relationship was noted between landscape broadleaf forest and avian population trends, such that the species most associated with broadleaf forests had greatest population declines. Therefore, maintaining or restoring broadleaf vegetation is important within coniferous dominated landscapes.

Because of landscape influences, providing desired forest conditions within forest stands may be insufficient for some priority species if these stands are located within inhospitable landscapes. Moreover, for long-term sustainable populations, most sites, even those with suitable habitat conditions, must have spatial and temporal linkages that facilitate dispersal, colonization, and interchange within metapopulations. Towards that end, desired landscape conditions for priority wildlife species are usually composed of large forest patches as the preponderance of the landscape matrix, have a wide range of variation in patch sizes and complexities, and provide extensive connectivity among forest patches.

Why Wildlife Forestry?

The application of wildlife forestry silviculture is in many respects the logical extension of the various alternative silviculture methods whose objectives are to maintain or increase forest biodiversity. Although forest management targeting increased biodiversity is a laudable objective in and of itself, managers of public conservation lands (e.g., National Wildlife Refuges and State Wildlife Management Areas) or private lands under conservation easements may have legal, or self-appointed, mandates to manage and conserve priority wildlife species rather than the more general goal of improving biodiversity. On other private lands, the economic and recreational values associated with silvicolous wildlife, including fee hunting, hunt-club memberships, wildlife viewing and other eco-tourism activities, have income potential that may equal or exceed that of extractive timber harvest. Hence, ensuring sustainable populations of wildlife may be essential for maintaining financial revenues from these forest lands.

In addition, land ownership patterns of private forest lands are changing. Changing ownership demographics and associated alteration of management objectives of these forest land owners have resulted in smaller landownership holdings. For owners of small landholdings, maintaining continuous forest cover and the aesthetic considerations associated with forests may be of paramount importance.

Table: State of habitat features following different distubances and prescribed wildife forsetry silviculture.

Legacy characteristic	Wildfire	Wind	Insect /disease	Flood	Clearcut	Wildlife forestry
Live trees	Sparse	Variable	Variable	Sparse	Sparse	Variable
Snags	Abundant	Variable	Abundant	Abundant	Sparse	Variable
Downed woody debris	Variable	Abundant	Variable	Variable		Variable
Understorydevelopment	Abundant	Abundant	Variable	Variable	Abundant	Variable
Spatialheterogeneity	Variable	Variable	High	Variable	Low	High
Duration of lterted state	Variable	Variable	Variable	Variable	Long	Variable

Even on large land ownerships, such as industrial forest lands or lands managed by timber consortiums, there are increased public expectations to provide sustainable, 'green' forest products. Concurrently, corporate efforts to maintain positive environmental images have prompted greater reliance on alternative silviculture, such as wildlife forestry. Finally, even on lands where implementation of traditional silviculture may be the preferred management option, exurban encroachment may constrain traditional forest management practices.

Priority Species

What are 'priority' species? For some groups of wildlife species, conservation planners have established priority rankings based on objective criteria such as current population abundance, distributions, availability of suitable habitats, perceived threats to habitat, etc. For example, for North American landbirds, Partners in Flight has established ecoregional priority rankings. For other wildlife groups, priority status may be conferred based on species being listed as threatened or endangered by national or regional governmental agencies or listed as a species of concern by private conservation organizations. Regardless of the underlying rationale for their selection, priority wildlife species within managed forests differ among regions and forest types. Moreover, the priority species targeted for habitat improvement via prescribed management actions ultimately are determined by the landowner's objectives.

Why favor these species over others? All forest management, including no active management (a.k.a., passive management or benign neglect), influences the abundance and composition of wildlife species. For many priority species, insufficient habitat, or an overabundance of habitat with unsuitable conditions, likely contributed to their designation as a priority species. Therefore, where modification of forest habitat will promote habitat conditions conducive to supporting sustainable populations of these species, it behooves managers to prescribe silvicultural actions that will result in increased habitat for these species.

Desired Habitat Conditions for Priority Species

Because priority species diffe r among regions and among forest types, quantification of specific habitat characteristics suitable for all forests and all species is not practical. Even so, empirical evaluations of wildlife species responses to different silvicultural treatments provide insight into the general habitat conditions that benefit different species groups.

Amphibians – Within landscapes, amphibians benefit from protection of existing forest, avoiding further forest fragmentation, and disturbances that mimic natural processes. Within stands, seasonally available water, especially vernal pools, should be protected from disturbance (e.g., siltation) by retaining forest canopy (≥75%) immediately surrounding pools and maintaining >50% canopy for up to 120 m from seasonal waters. Amphibians tend to benefit from retention of sufficient canopy throughout the forest stand to provide a partially shaded forest floor. Their abundance also tends to be positively associated with a deep litter layer and abundant woody debris on the forest floor. Within southern pine forests (e.g., longleaf pine [Pinus palustris]), thinning and burning are positively related to increased abundance of priority (open-pine adapted) amphibian species.

Reptiles - The effects of silvicultural treatments on reptiles have been mixed, often exhibiting no discernable effect. Despite inconclusive studies, development of heterogeneous canopies, with numerous gaps enabling light to reach the forest floor, enhances thermoregulation opportunities for reptiles. Herpetofauna (reptiles and amphibians) may exhibit different regional responses to silviculture. Similarly, species-specific response to silviculture may result in species responding positively, negatively, or not at all. Thus, more research is needed on the effects of thinning and retained structural elements on reptiles and amphibians.

Birds: The intensity of harvest, otherwise stated as the magnitude of retention, influences species composition of bird communities. More canopy retention favors retention of mature- forest species, whereas more intense harvest tends to favor early-successional species and forest generalist species. In northern hardwood forests, retention of, 20 m2 ha-1 basal area maintained habitat for mature-forest birds yet afforded habitat for birds using early-successional habitat. Similarly, in South American hardwood forests, combined dispersed and aggregated retention permitted colonization by early successional birds yet retained species characteristic of old-growth forests within retained aggregates. In Tasmanian eucalypt forest, long-term, aggregated retention harvests better sustained mature forest bird communities, post-harvest, compared with clearcut or dispersed retention harvests. In young, conifer- dominated stands, thinning promotes diversity of breeding birds – a variety of thinning intensities and patterns, ranging from no thinning to widely spaced residual trees, will maximize avian diversity at the landscape scale and provide structural diversity within and among stands. In southern pine forests, bird abundance and richness were positively related to volume of coarse woody debris and density of snags. In Midwestern oak-hickory forests, thinning harvest and prescribed fire increased avian nest survival (Streby and Miles, 2010). However, responses to variable retention harvests are species specific and temporally dynamic: Response of some species is immediate but short- lived, whereas other species exhibit a deferred (maybe several years post-treatment) but long- lasting respons.

Bats – Many bat species avoid spatial clutter: Thus bats tend to be more abundant in forests with open structure (harvested or old-growth) compared to closed-structured mid- successional forest.

Forest structure appears more important than forest composition for foraging bats. Increased species and structural diversity of woody species, including providing broadleaf trees and shrubs in corridors and surrounding water features, tend to enhance foraging opportunities for bats. Conversely, forest type (or tree species) may be an important determinant of roosting location or behavior. Specifically, roost trees of bats are generally tall with large diameters and are located in stands with open canopies and high snag densities. Compared to trees used by foliage-roosting bats, cavity-roosting bats use trees within stands that have more open canopies and trees that are closer to water than are random trees.

Deer – Within landscapes, decreased road density and lessening of associated traffic reduces the negative impacts of disturbance. Silviculture should promote structurally complex, uneven-aged stands, that include forest openings (1 – 20 ha), and provide structural heterogeneity within and among stands. Resulting habitat should provide understory forage and cover while retaining species diversity of trees.

Bear – Habitat selection is largely influenced by food abundance. A diversity of tree and shrub species producing hard mast (acorns and nuts) as well as soft mast (berries and fruits) is beneficial. Thinning, group selection, or patch cuts should be used to promote regeneration and recruitment of mast producing species, as well to increase understory food production, escape cover, and bedding areas. Care should be exercised to retain some large trees within each stand and to protect potential den trees during harvest.

Small Mammals –Silviculture that retains legacy structures, variable-density thinning, and management for increased forest decadence combined to support 3 species of squirrel in Douglas-fir forests. In southwestern conifer forests, thinning and burning treatments had positive effects on most small mammals. Even so, effects of silviculture on small mammals are species-specific, such that a positive response in abundance of one species may be offset by negative response in abundance of another species.

Effective strategies to achieve habitat for different species across spatial and temporal scales should include developing:

- Measurable objectives.
- Integrated conservation goals and silvicultural prescriptions.
- Clear and practical guidelines, effective training, and communication programs.
- A monitoring and an adaptive management process to evaluate and improve results.

Social Forestry

When the forests are managed with benefits of the whole community in mind, that type of forestry is called social forestry. The aspects included in social forestry are forest management, afforestation of deforested areas and existing forest protection; having the objective to improve the rural, environmental and social development. When it comes to social forestry, the needs of the local communities come first.

This is the reason due to which the goal of social forestry is to grow a lot of trees and plants to make sure they meet the growing needs of the people for their demand of timber, food, fuel, wood and dependency on traditional forest areas. Social forestry also protects agricultural areas from adverse climatic conditions to improve the environment, increase the supply of forest/agricultural produce and to enhance the natural beauty of the local area.

This concept and practice has existed for many centuries, gaining new dimensions constantly as it also helps to tackle the challenges created by global warming.

Major Types of Social Forestry

Agroforestry

This involves growing trees in a way to provide the landowners with forest and agricultural products on a commercial basis. The main objective of this is to have positive interactions between the two systems of forests and agriculture. These systems can also be separately practiced.

This is the best type of social forestry if someone wants to start a venture of farm forestry while maintaining the existing agriculture practices. This type of forestry provides economic as well as social benefits. Additionally, this goes perfectly along with one of the major principles of Permaculture. Permaculture is one of the Sustainable Farming methods.

Farm Forestry

In this type of forestry, the objective is to grow trees for a specific purpose of harvesting a product. The main purpose is usually to plant timber trees on private lands, but this venture can also be used in other ways, like for fruit farming.

Farm forestry has many benefits which include shelter and pasture for animals; diversification of earnings, increase in capital value, improved environmental conditions, improved living and heightened biodiversity.

Extension Forestry

This type of forestry is becoming very common in urban cities as it involves the planting of trees on the sides of roads, canals and railways. This is a very viable option for local governments looking for betterment of their communities.

Community Forestry

This refers to the management of the communal land. The projects are decided and implemented by the village members on the communal land. The local people take part in managing, planning and harvesting the crops. They even share the socio-economic and ecological benefits. The main purpose of such forestry is to increase the involvement for local people and lead the whole village into prosperity. Another major advantage is that there is reduced inequality between members of the community/village which is sometimes the root to evil actions.

Benefits of Social Forestry

Increased Biodiversity

When trees are grown in barren lands, they help to increase the biodiversity value among the community on a whole other level. When trees grow larger and longer they enhance the nature of habitat. Forests which are well maintained and managed properly, they encourage biodiversity as they help to offer habitation for various animals, shrubs, plants, birds and insects among others. In nature, wherever and wherever trees grow; other plants and wildlife follow.

Trees and many other plans become the source of food and shelter for many small animals and birds. Many mature trees create an environment which encourages the growth of many other plants which would not have existed otherwise. This therefore increases the food varieties for animals and the local people living nearby.

Carbon Removal

Trees act as carbon sinks. Trees play a very important role in the fight against global warming, as they remove carbon from the air. While growing the trees absorb carbon dioxide while removing it from the environment. Majority of the carbon emissions in urban cities come from cars, trains and whatnot; social forestry is seen as the best way to counter it.

If 100 million trees are planted, they would save 22 billion kilowatt hours(KWh) of energy and about 3 million tons of carbon emissions for carbon dioxide annually in a time span of ten years. However, it should be understood that the actual amount of carbon dioxide which would be removed would depend upon the types of trees and vegetation planted in the forest.

The places which are well covered by trees can be up to 6-7 degrees cooler than the areas which are devoid of trees.

Soil Conservation

This is also one of the major benefits of social forestry. Communities which embrace social forestry have had many significant economic and health benefits which they enjoy in terms of improved agricultural activities. Tree roots hold onto the soil which protects the soil from erosion, keeping it in place. In a medium-sized city, the trees which are planted along the roads and in parks can help save around 10 tons of soil annually. The tree leaves which decay also help the soil to get rich as they form an organic layer on the ground allowing the water to percolate onto the soil (A technique used in organic and sustainable farming methods). This reduces the chances of soil erosion.

Trees greatly reduce the likelihood of overland flow of water as the tree roots store the water among them and hold the soil in place too!

Health Benefits

There are many researches on how beneficial trees are for humans. Trees and other aspects of nature are all natural remedies for anxiety and stress. Taking a walk in the park has been a great source of calm for people when they are stressed. Planting trees in human habitat would contribute a lot towards better health and improved well-being. It is now very common that hospitals are planting and growing trees because of their healing effects. Trees also help in absorbing many pollutants from the air that may have negative effects on human health. They even provide medicinal value for numerous communities and healthcare providers.

Community Enrichment

Trees help to make the neighborhood look more serene and attractive. Community Areas with trees attracts more people for living rather than an area which does not have trees (this also has an economically beneficial side to it). Trees also provide shade which improves the environment surrounding your outdoor furniture.

Areas which are well stocked with trees help keep the community together as the inhabitants have a common goal of betterment of community.

Noise Reduction

Noise triggers anxiety and stress. Loud noises can disturb sleep, affect behavior of people and even cause mental illnesses. Being exposed to loud noise over a long period of time can even cause hearing loss. Many solutions are being employed to reduce noise pollution and among those is social forestry as it helps to reduce noise pollution. Trees absorb and reflect sound energy which reduces noise pollution.

Improved Air Quality

Most people associate trees with just the removal of carbon dioxide, that can never be overlooked but trees also help to improve air quality and clean surrounding air. An acre of trees help to generate

enough oxygen for 18 people. Trees also absorb pollutants which are harmful to people's health. Polluted air is often a major reason for exacerbation of asthma or COPD.

Basically, trees clean the air by absorbing all the pollutants from the air, as their leaves store and trap the particles which get filtered through their stems, leaves and twigs. The smallest particles trapped by plants are around 10 microns small.

Energy conservation and reducing overall atmospheric temperatures. Trees are natural air conditioners. The evaporation that can happen from a single tree can have the cooling effect similar to ten room-sized air conditioners operating together almost throughout the day. Planting trees at your home would help reduce the outdoor heat around the house and the air-conditioners would have to work less which would save energy and subsequently lower your bills.

Social Benefits

Trees have many social benefits which makes the neighborhood more valuable and attractive. Social forestry also helps with creating more inviting and cooler areas for relaxation such as playgrounds and parks. They even add up to color and beauty.

The green color of the trees creates an effect that is calming and helps the eye to be relieved of strains. Trees also provide a shaded place where people can meet under the scorching sun.

References

- Social-forestry-types-and-benefits, plants-and-trees: environmentbuddy.com, Retrieved 21, June 2020
- Energy-forestry-41155: hisour.com, Retrieved 15, March 2020
- Forestry-definition-branches-costs-programme-and-conclusion-geography, forestry-8370: geographynotes.com, Retrieved 09, August 2020
- Sustainable-forestry: extension.psu.edu, Retrieved 27, July 2020
- Ecoforestry-ecoplan-forestry: ecoplan.ie, Retrieved 16, May 2020

Chapter 2

Forest: Conservation and Management

The practice of planning and maintaining forested areas for the sustainability of future generations is known as forest conservation. The branch of forestry deals with the administrative, social, economic, scientific and technical aspects is referred to as forest management. This chapter provides comprehensive information regarding forests and their conservation and management methods.

The forest is a complex ecosystem consisting mainly of trees that buffer the earth and support a myriad of life forms. The trees help create a special environment which, in turn, affects the kinds of animals and plants that can exist in the forest. Trees are an important component of the environment. They clean the air, cool it on hot days, conserve heat at night, and act as excellent sound absorbers.

Plants provide a protective canopy that lessens the impact of raindrops on the soil, thereby reducing soil erosion. The layer of leaves that fall around the tree prevents runoff and allows the water to percolate into the soil. Roots help to hold the soil in place. Dead plants decompose to form humus, organic matter that holds the water and provides nutrients to the soil.

Plants provide habitat to different types of organisms. Birds build their nests on the branches of trees, animals and birds live in the hollows, insects and other organisms live in various parts of the plant. They produce large quantities of oxygen and take in carbon dioxide. Transpiration from the forests affects the relative humidity and precipitation in a place.

The FAO (Food and Agriculture Organization) has defined forest as land with tree crown cover (or equivalent stocking level) of more than 10% and area of more than 0.5 hectare. The trees should be able to reach a minimum height of 5 m at maturity in situ. Forests are further subdivided into plantations and natural forests. Natural forests are forests composed mainly of indigenous trees not deliberately planted. Plantations are forest stands established by planting or seeding, or both, in the process of afforestation or reforestation.

Forests can develop wherever the average temperature is greater then 10 °C in the warmest month and rainfall exceeds 200 mm annually. In any area having conditions above this range there exists a variety of tree species grouped into a number of forest types that are determined by the specific conditions of the environment there, including the climate, soil, geology, and biotic activity.

Forests can be broadly classified into types such as the taiga (consisting of pines, spruce, etc.), the mixed temperate forests (with both coniferous and deciduous trees), the temperate forests, the sub tropical forests, the tropical forests, and the equatorial rainforests. The six major groups of forest are moist tropical, dry tropical, montane sub tropical, montane temperate, sub alpine, and alpine. These are subdivided into 16 major types of forests.

Importance of Forests

Forests host 80 % of all terrestrial biodiversity. All these animals, plants, fungi and bacteria provide essential services for nature and the survival of mankind. They also hold enormous economic significance.

Forests are Vital to our Health

As a guardian of terrestrial biodiversity, forests are a treasury of medicinal plants and pharmaceutical ingredients. Some 75 % of the most commonly prescribed medicines source their ingredients from forests. Not to mention the healing effects of clean water and air and a healthy walk in the woods.

Fig: An okapi in a forest.

Forests give us Oxygen

A single mature deciduous tree generates enough oxygen for two to ten people. Without oxygen, life as we know it would not be possible. Although the phytoplankton in the oceans accounts for at least half of the world's oxygen, forests thus also play a key role.

Forests Account for 75 % of Freshwater Resources Worldwide

Water is better absorbed into the soil thanks to trees and forests which improve groundwater recharge and feed springs and rivers. At high altitudes, cloud forests can extract moisture from fog and clouds.

Forests Regulate Precipitation Worldwide

Normal rainfall patterns are heavily dependent on forests, not only because they massively evaporate water, but also because they emit biological particles such as pollen and fungal spores, which can serve as the nuclei of rain droplets or snowflakes. Coastal forests draw in moist air from neighboring seas and carry it all the way into the drier inland regions, the effects of which can be felt over thousands of kilometers. Forests are therefore essential in times of climate change, when longer periods of drought occur more frequently.

Fig: Three persons from an indigenous tribe with their faces painted.

Trees and Forests are Natural Cooling Systems

Under the crests of trees and forests, the temperature is noticeably lower than in open areas or in the city, as trees use solar energy to evaporate moisture, which has a cooling effect. A single tree displays a cooling effect similar to the aircons of two average households. As a result, the surface temperature of forests can be up to 20 degrees cooler than that of buildings or road asphalt.

Forests Slow Down Global Warming

After oceans, forests are the world's largest carbon sink. More forests means less CO2 in the air and less global warming. This comes on top of the natural cooling effect of forests. Approximately 18 % of current global warming is due to loss of forests.

Fig: A mangrove forest in Thailand.

Forests Protect us from Disasters

Floods are less common in areas where forests grow. One reason for this is that, during heavy rainfall, much more water seeps into the soil, instead of flowing directly into the rivers. Mangrove forests protect the coast from tsunamis. Forests also reduce the risk of landslides, avalanches and sand storms.

Forests and Trees Grow on more Fertile Soils

Thanks to forests and trees, fewer fertile soil particles flow away during precipitation. The roots of trees bring nutrients from the deep soil to the surface, while fallen leaves feed the soil and ground-dwelling organisms.

Fig: Fruits from the wild rose (rosehips).

Forests Purify Soil, Water and Air

Not only houseplants, but also forests purify the air of pollutants, including carbon monoxide, sulphur dioxide, nitrogen dioxide and fine dust. They can also help purify contaminated soils.

Forests Contribute to our Well-being

Green makes you happy, and forests even more so. Taking in the atmosphere of a forest calms us down.

Forests Provide Jobs and Incomes

The economic value of the services provided by forests worldwide is estimated at 16.2 trillion dollars. In addition to the supply of clean water and air, these services include numerous indispensable products, including wood (construction, furniture, paper, energy, etc.), resins, fruits, honey, ingredients for medicines, cosmetics, detergents to name just a few. They also have a role to play in tourism. On top of that, forests are said to create jobs for more than 13 million people. It is also important to note that about a third of the world's population still depends on forests and trees for their daily needs, especially for cooking and heating.

Types of Forest

Most of the animals are living in the forests and it is important for us to survive. Forests are helpful for human also and can be available in any of the places in whole of the world.

Types of Forests

Tropical Evergreen Forests

- Location: The tropical rainforests contain the greatest diversity of species of all biomes on the ground. They are located around the equator, between 23.5 degrees N latitude and 23.5 degrees S latitude.
- Climate: Temperatures in tropical rainforests are between 68 and 77 degrees Fahrenheit annually. Winter is missing in these forests. Most tropical rainforests receive 100 inches of rain a day.
- Soil: Because the temperature is warm and the air is humid, the decomposition is very rapidly occurring in the tropical rainforest. Advanced precipitations often lead to the nutrition of nutrients from the soil, which causes poor food-poor soil.
- Plants: Trees in the tropical rainforest grow between 82 and 115 meters high and are typically broad leaves. Other plants include children, vines, mice, palms, and orchids.
- Animals: Dense growing trees create a thick canopy of tropical rainforests, which hold the sun to penetrate into the low streets of the forest. This means that most animals living here must adapt to life in the trees. A variety of birds, bats, monkeys, serpents and other animals are found in tropical rainforests.
- Threats: We can find the biggest threat to the tropical rainforests was forestry practices.

Temperate Deciduous Forests

- Location: The Eastern United States and Canada, Western Europe and parts of Russia, China, and Japan.
- Climate: There are four different seasons in temperate deciduous forests and precipitations fall during the rain during rain, summer and autumn and snow in winter. Temperate deciduous trees receive 30-60 inches of rain.
- Soil: The soil in these forests is very prolific.
- Plants: Forests in moderate deciduous forests support mice, children and wildflowers and support varieties of shrubs and ferns. The trees that dominate this forest are consists of Oaks and birches which are the examples of these forests and a small number of the evergreen forests consists of pines and spruce.

Temperature Coniferous Forests

- Location: Coniferous forests are those forests which presents in the coastal areas with warm winters and heavy storms. For example, the Pacific Northwest of the United States and Canada, Southwest America, Southern Japan, New Zealand and small parts of North-western Europe (Ireland, Scotland, Iceland and Norway).

- Climate: The warm weather is a low temperature throughout the year. High precipitation levels (50 – 200 inches per year) produce a humid climate and a long period of time.
- Soils: Soils are usually rich in a thick layer of material deterioration.
- Plants: Perennial conifers dominate these forests. Due to the high levels of precipitation and moderate temperatures, there is a long-term season due to the high growth tree. The temperatures of pine trees are the most important trees with cedar, cypress, Douglas fir, pine, adventures, and bushes. There are tree leaves, such as maple trees, and mosses and ferns are common.
- Animals: Examples of animals living in temperate coniferous forests, deer, marmots, hawks, black bears, salmon, white owls, marble marten.
- Threats: Sustainable forests, road construction, and other development activities pose the greatest threat to the deciduous forest.

Taiga Forests

- Location: This is the type of northern forest that lies between 50 and 60 degrees latitude. The boreal forests are in Canada, North Asia, Siberia and Scandinavia (Denmark, Norway, Sweden, and Finland). Two-thirds of the global indicators were found in Scandinavia.
- Climate: Boreal forests are long winters and short summers. Most precipitation is snowy and these forests are annual rainfall ranging from 15 to 40 cm.
- Flooring: At low temperatures decomposition takes a long time, because of the fine earth.
- Plants: The trees are perennial leaves, such as beef, spruce, and pine. The ground floor is limited because the glass is so compact.
- Animals: Animals that are here must adapt to cold and cold winters and generally have thicker skins. Deer, elk, holm oak, caribou, snowshoes, wolves, grizzly bears, lynx, and goblins are some examples.

Forest Conservation

Where geographical conditions permit vegetation to take the form of trees, the forest is one of the major forms of the natural landscape. The forest resources are valuable as an integral part of the ecosystem, from the commercial point of view, and as providers of shelter to wildlife. Today forests provide the raw materials for over 5,000 products worth about 23 million dollars.

They support industry which employs 1.3 million people. In fact, forests are still the natural habitats of several species of plants and animals, as well as of several tribal groups of the world. But, the most unfortunate setback came in the form of commercial exploitation, which resulted in mass destruction of forest cover year after year.

Originally, over two-fifth of the land area of the earth, exclusive of the Polar Regions, or about 1,200 million hectares was covered with natural forests. But, now more than one-third of this area has been robbed by man of its natural protective cover and has been turned into barren land.

The history of the exploitation of forests is as old as man himself, but during earlier times it was balanced through a natural growth process because at that time forest cutting was done for personal or community use only. But with the expansion of agriculture, forest lands have been cleared.

More destruction has been done after industrial revolution and urbanization. During the colonial period commercial exploitation began and this was the main cause of the depletion of forests. The commercial use of forests nowadays has reached such an extent that it has become a threat to the environment in the form of:

- Increase in temperature.
- Lesser precipitation.
- Increased rate of soil erosion.
- Increase in frequency and volume of floods.
- Loss of soil productivity.
- Extinction of several species.
- Non-availability of several essential forest products.
- Imbalance in ecosystem.

The harmful effects of deforestation are so much that all over the world people and authorities have realized that forest resources must be conserved properly in order to protect the ecosystem. The forest is a national resource and a social asset. It yields a great social profit which lies wholly outside the realm of business. But, at present, most of the forests of the world are so over-used that experts predict dire calamities in the not-too-distant future and irreparable damage on a catastrophic scale. If properly used and put on a sustained yield basis, it will be one of man's greatest resources and for this; conservation of forest is the only alternative.

Methods of Forest Conservation

The following steps should be taken for the conservation of forests:

Regulated and Planned Cutting of Trees

One of the main reasons of deforestation is commercial felling of trees. According to an estimate, about 1,600 million cubic meters of wood have been used for various purposes in the world.

Although trees are considered as perennial resource, when exploited on a very large scale, their revival cannot be possible. Therefore, cutting should be regulated by adopting methods like:

- Clear cutting.
- Selective cutting.
- Shelter wood cutting.

The clear cutting method is useful for those areas where the same types of trees are available over a large area. In that case, trees of same age group can be cut down in a selected area and then marked for replantation.

In selective cutting only mature trees are selected for cutting. This process is to be followed in rotation. Shelter wood cutting is where first of all useless trees having been cut down followed by medium and best quality timber trees.

The time gap between these cuttings is helpful in re-growth of trees. In regulated cutting only one-tenth of the forest area is selected for use and rotational system is always followed for their protection. The forest can be managed in such a way that a timber crop may be harvested indefinitely year after year without being depleted. This technique is called the 'sustained yield' method adopted by many countries of the world.

Control over Forest Fire

Destruction or loss of forest by fire is fairly common; because trees are highly exposed to fire and once started it becomes difficult to control. Sometimes, the fire starts by natural process, i.e., by lightning or by friction between trees during speedy winds, while in most cases it is started by man either intentionally or unintentionally.

According to an estimate, during the period from 1940 to 1950, in the US alone, fires consumed an average of 21.5 million acres of timber yearly and as many as 1,175,664 cases of forest fires occurred during 1955 to 1964 period and a spark either from a lightning bolt or a match in the hands of a fool or a knave. The formula is simple – the larger the forest, the drier the air, the bigger the fool, the bigger the fire you will have".

In order to save forests from fire it is necessary to adopt latest techniques of firefighting. Some of the fire suppression techniques are to develop three metre wide five lanes around the periphery of the fire, back fires, arrangement of water spray, fire retardant chemicals should be sprayed from back tank and if possible by helicopters. There must be a trained staff of firefighters to control the fire.

Reforestation and Afforestation

The sustained yield concept dictates that whenever timber is removed, either by block cutting or by selective cutting, the denuded area must be reforested. This may be done by natural or artificial methods. Similarly, any forested land which has been destroyed by fire or mining activities should be reforested. In rugged terrain aerial seeding is the method of choice.

Besides all this, fresh afforestation programmes should be started. New plantations will not only increase the forest cover but also help in making up the eco-balance. For afforestation, selection of trees should be done according to local geographical conditions and care must be taken during initial growth of the trees.

Check over Forest Clearance for Agricultural and Flabitation Purposes

Most of the present-day agricultural land was once forested and then cleared for the use of agriculture. But now it has reached the stage where further clearance will be dangerous for the entire ecosystem.

There are tribal in some parts of Asia, Africa and South America, where shifting cultivation is still a part of their system of land procurement. According to an estimate, about 40 million sq km of land is used for this purpose by 200 million tribal of the world.

For the conservation of forest, this should be checked and an alternative system should be suggested to them. Similarly, for the development of villages, towns and cities, forest lands have been cleared and this process continues to this day causing loss of forest cover. This also should be checked and green belts around cities should be developed.

Protection of Forest

The existing forests should be protected. Apart from commercial cutting, unorganised grazing is also one of the reasons. There are several forest diseases resulting from parasitic fungi, rusts, mistletoes, viruses and nematodes which cause the destruction of trees. The forests should be protected either by use of chemical spray, antibiotics or by development of disease resistant strains of trees.

Proper Utilization of Forest Products and Forests

Generally, forests have been cut for logs and rest of the tree – stump, limbs, branches and foliage, etc., are left out in the forest as worthless debris. Further waste occurs at the sawmill. There is a need to use all this waste material. Now several uses have been developed and products like waterproof glues, board, etc., can be obtained.

Similarly, forests can easily be used or developed as tourist centres. By using them as tourist centres the country can earn substantial foreign exchange. This practice has been adopted by many countries, both developed and developing. The concepts of 'national park' and 'game sanctuary' have now become popular and every country has developed its unique forest area as a 'national park'.

Forest Management

Forest Management is defined in the Glossary of Technical Terms as the practical application of the scientific, technical and economic principles of forestry (BCFT). The term is variously defined by different authors, embodying, in essence, the same essential ingredients.

Some of these are reproduced below:

- "Forest Management is that branch of forestry whose function is the organization of a forest property of management and maintenance, by ordering in time and places the various operations necessary for the conservation, protection and improvement of the forest on the one hand, and the controlled harvesting of the forest on the other."
- "Forest Management is the application of business methods and technical forestry principles to the operation of a forest property" (SAF).

The above definitions highlight the varied nature of the subject which is concerned with the task of "building up, putting in order and keeping in order a forest business". Forest management, by implication, is not a basic subject in itself; it is the practical application of science, technology and economics to a forest estate for the achievement of certain objectives – mainly production of wood – timber and industrial raw material, and other forest products such as resin, gum, tan bark, etc. It is based on the knowledge of a number of basic subjects sciences, such as Siliviculture, Ecology, Geology, Pedology, Botany, Mensuration, Pathology, Economics, and Finance etc. In addition, a forester needs the practical experience gained from observations in the field, results of past treatments given to a forest and deductions there from.

Management of forests broadly involves three main tasks via:

- Control of composition and structure of the growing stock.
- Harvesting and marketing of forest produce.
- Administration of forest property and personnel.

It is, unlike any other commercial enterprise, complication; as forests, particularly the State-owned as most of the forests are in India (95.8%), are managed for a multiplicity of purpose – productive, protective, climatic, wildlife, recreational and bilaesthetic, with one use dominant, viz., most often the production of wood. Though the forest land can be managed simultaneously for several uses, however, in some cases, uses are incompatible with one another; e.g., grazing is not compatible with timber production, environmental conservation and recreational use. In each case, priorities are laid down by the owner - the State or the private owner as the case may be, and the management is oriented to achieve the prescribed objectives. In the forests dedicated primarily to recreational and bio-aesthetic use, and conservation of ecology and environment, grazing, fellings, timber extraction and even hunting may have to the suspended. Productive and protective functions of the forests cannot be bifurcated. As a matter of fact, scientifically managed forests perform both these, simultaneously.

Productive forests do protect and protective forests do produce – the distinction between the two is of degree rather of kind - a matter of emphasis of the primary function of the forest that the management aims at. It is, therefore, essential that forest resources are maintained in a state of maximum production, consistent with their subsidiary or even the other main functions. Forests have to be managed in such a way so as to provide maximum benefits to maximum people and for all time, ensuring that the soil produces most and deteriorates least under their treatment.

Scope of Forest Management

Management of Forests, as that of any other enterprise, involves a process of making and implementing policy decisions to achieve the objectives of the owner. These decisions involve, in turn, a plan of action. Planning is the responsibility of the States and the Centre in case of State-owned forests, broad principles for which are embodied in the National and the State Forest Policies. Detailed plans are prepared by Forest Managers at the professional level, and executed by the technician level staff. Forest Manager has to constantly manage the growing stock to achieve given objects of management; in this process he has to decide: 'how much, when where and how to cut.' Scope of Forest Management is very extensive; it encompasses broadly, the following main activities:

A. Control of Growing Stock, its Structure and Composition:

- Site adaptation.
- Choice of species.
- Manipulation of stands.
- Harvesting the produce.
- Regeneration.
- Protection.

B. Distribution and Marketing of produce:

- Transportation and communication.
- Logging Plan.
- Marketing data.
- Sale of produce.
- Revenue.

C. Administration of Forest Property:

- Forest organization.
- Management of Personnel.
- Monitoring and control or works.
- Labour management and welfare.
- Financial control and economy efficiency.
- Fulfillment of social obligations.
- Record for present and future reference.

Forest Management Planning

Planning is an active process requiring careful thought about what could or should happen in the future and involves the coordination of all relevant activities for the purpose of achieving specified goals and objectives. Planning is an integral component of forest management; is about determining and expressing the goals and objectives which government, rural communities or companies have, and for deciding the targets and steps that should be taken in order to achieve those objectives.

Planning need not be a complicated process but it requires clear objectives which a government or other group aims to achieve. It requires imagination and a willingness to consider all points of view having relevance to a given situation. The planning process should lead to the formation of a balanced outlook from which proposals for effective management can be written. An element of flexibility is desirable and necessary, however, in order to cope with unforeseen events which could affect the achievement of the objectives.

A range of information is used in planning to evaluate the benefits and drawbacks of alternative courses of action, which enables preferred options to be determined, coordinated with other activities, and expressed in writing. Information should be of good quality. Information of questionable quality should either be discarded or, if used, it should be noted that it is of poor quality and one of the activities of the plan should be acquiring better quality information.

Two Time Horizons of Planning

Forest management plans should have a minimum duration, or length, of 10 years. A shorter period than 10 years does not provide the medium-term stability that is needed to guide consistent implementation of sustainable forest management activities. A realistic maximum length is 20 years. The duration is also called the term, or period, of a plan. A management plan should include prescriptions that provide for:

- Review at the mid-point of the plan,
- Review in the final year of the plan,
- The preparation of a new plan upon expiry of the present plan.

Annual plans of operations are written for a one-year period and should be derived from a five-year (or longer) management plan. A Plan of Operations should express specific activities in tabular form for one year only and for a specific locality, such as a felling area.

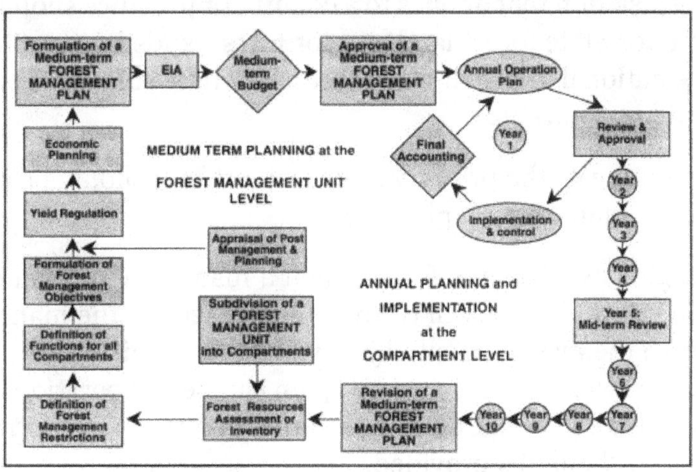

Fig: The Forest Management Cycle.

The relationships between elements of medium-term and annual forest management planning cycles are shown in figure. Table below describes a tiered structure of long-, medium- and short-term management relationships. A number of commonly used terms in forest management planning and in the implementation of plans are defined in Figure Commonly Used Terms in Forest Management Planning.

Table: The Tiered Forest Management Planning System.

Management Level		Time Frame	Key Elements of Forest Management
Strategic Issues	National	> 10 years	Conservation driven forest policies
	Sub-national	> 10 years	Forest Sector Development Plans
Operational Issues	Forest Management Unit	10 years	1. Management function - Planning - Integrated forest resources inventory - Environmental impact assessment - Forest zoning by forest functions - Forest protection - Integration of customary land rights - Yield regulation
	Compartment	Annual	2. Management Function - Implementation - Low impact forest resources management - Yield allocation according to stand maturity 3. Management Function - Control - Forest resources accounting - Periodic revision of management plans

Legislation for Management Planning

A component of forest legislation that applies to a country or province should be that management plans are to be prepared for State forest land and for forest lands in non-State tenure the conservation of which is in the national or provincial interest or where subsidiaries or incentives are paid to promote forestry development.

The following guidelines indicate the primary requirements that should be included in forests legislation in respect of forest management planning:

- Subject to the rights existing in a forest when a management plan comes into operation, forests legislation should specify that a plan must regulate the management of forest land in conformity with the management objectives for a specified time period.. The maximum number of years for which a management plan can be in operation should be specified. It should not be less than five year nor more than 20 years. Scope for review of a plan during the planning period should be provided.

Forest Management Plan: *A document that translates forest policies into a coordinated programme for a forest management unit and for regulating production, environmental and social activities for a set period of time through the use of prescriptions specifying targets, action and control arrangements.*
Forest Management Unit: *An area of forest for which an approved. Forest Management Plan is in operation, or any subdivision of it, such as a block (a specified locality) or a felling series.*
Felling Series: *Specific compartments of a Forest Management Unit that are being managed in a specific way for silvicultural reasons and also have an independently calculated harvest yield.*
Plan of Operations: *A tabular statement showing the order and extent of forest work to be carried out during one year and is based upon the prescriptions of a Forest Management Plan. The term is synonymous with Annual Work Plan.*
Compartment: *A permanent, geographically recognisable unit of forest land forming the basis of prescription and permanent record of all forest operations.*
Sub-compartment: *A sub-division of a compartment, sometimes temporary, where because of particular characteristics an area of forest has a special description and requires separate silvicultural treatment.*
Prescription: *A concisely written specification in a forest management plan that translates an objective, or a part of one, into an operational activity.*
Mean Annual Increment (MAD): *The total increment, usually volume, up to a specified age, divided by that age.*
Current Annual Increment (CAI): *The average increment, usually volume, for a management plan prescription, including where, when and how the yield should be extracted.*
Yield Regulation: *The determination of the yield and its expression in a management plan prescription, including where, when and how the yield should be extracted.*
Working Circle: *One or more parts of a working plan area, not necessarily adjoining, having the same objectives, silvicultural system and prescriptions.*
Working Plan: *A document that regulates wood production, silviculture and tending activities for a specific period of time through the application of prescriptions specifying targets, action and control arrangements. A Working Plan is not synonymous with a Forest Management Plan.*
Source: Terminology of Forest Science. Technology, Practice and Products. Society of American Foresters. 1971.

Fig: Commonly Used Terms in Forest Management Planning.

Each forest management plan should specify:

- The maximum area from which forest produce may be harvested, or the maximum quantity of forest produce which may be harvested, or both, in a given time period.
- The forest protection operations to be carried out.
- The forest development operations to be carried out, including silviculture.
- Other matters which are necessary or appropriate in order to implement management objectives effectively. This could include forest inventory, mapping, technical and social surveys, and public consultation. A management plan to be applied to State or private forest land should be approved by the Ministry responsible for forestry or other specified authority.

Balance between Production, Social and Environmental Objectives

Forests provide a wide range of benefits at local and national levels. Log production is usually the main objective and revenue earned for governments and companies is the major driving force in tropical forest harvesting. Revenue earned from log harvesting will usually be the main funding source for long-term sustainable tropical forest management.

Many communities depend heavily upon non-wood forest products for subsistence and as a basis for local trade, for example, canes, medicinal and food plants, gums, resins and wildlife. Tropical forests are an essential source of energy for many communities, directly through burning of wood for cooking and heating and indirectly to protect watersheds as source of water for hydroelectricity generation. They have an important role in protecting physical and biological environments at local and provincial levels. Tropical forests are dwelling places for many millions of people and are increasingly of value for recreation and tourism, notably "eco-tourism". They are important havens for wildlife and are the habitats of many endangered species of plants and animals. It is essential in tropical forest management planning to achieve a balance between the long-term between wood production, social and environmental management objectives.

The Participation of all Interest Groups in Planning

The initiative for the formulation of a forest management plan should be taken by the forest owner, such as the state, or by a concession holder on behalf of the owner. Local communities and others having historical rights or privileges in a forest are important stakeholders and must be involved in planning. The likelihood that sustainability will be achieved will be considerably enhanced if local people who live in and around tropical forests have a say in and participate in management planning and are able to share in the benefits of forest use to ensure that their basic needs are met. The steps that can be taken can include:

- Granting secure tenure to existing productive farmland within the forest.
- Local participation in management decision-making.
- Guaranteed access to forest products.
- Provision of employment.
- Shared benefits in forest harvesting.

Accommodation of the respective interests of forest management companies and local community groups can be eased with the recognition that companies and indigenous communities are, in most cases, interested in different things - industrial logs by forest management companies on the one hand and small dimension wood and non-wood products by local communities on the other. A positive approach towards adapting the interests of both groups is to include rural communities as partners in forest management and to share the benefits of wood production with them.

Yield Prediction

Types of Yield Prediction Models

A yield prediction model uses the quantitative relationships between measured growth variables to predict yields of forest types, and is a tool that helps to schedule and regulate harvests at sustainable levels. Two basic methods are available for their construction, diameter class (or stand table) projection and cohort modelling. Both depend upon the use of comprehensive growth data to construct and fit a yield prediction equation.

The Basic Steps Involved in the Construction of a Yield Prediction Model

Diameter Class Growth Projection

The manual method of diameter class growth projection is the oldest method, first used in Myanmar in 1856, and used elsewhere for simulating the growth of tropical forests. A tree population/ diameter class distribution is compiled using PSP summary data. DBH classes are in 5 or 10 cm intervals. Stratification of the tree population can be made on the basis of species groups. Average diameter growth and average tree mortality rates are determined from periodic measurements of PSP, for each diameter class. Growth is projected for a five-year period for each diameter class per hectare by applying growth and mortality rates.

Construction of a Growth and Wood Yield Prediction Model

The process of fitting a yield prediction model from forest data may be by fitting the field data to a pre-determined regression, for example, a linear regression, or manually by plotting the data on graph paper; equations can then be developed from the hand-drawn curves. The use of personal computers allows statistical procedures to be applied with greater precision and more quickly than manual methods.

Testing of a Yield Prediction Model for its Validity

A wood yield prediction model must be tested to determine its validity and precision. The precision of a yield model will depend on how well the PSP represent the forest, the number and period of the remeasurements, on covariances of the predictor variables and on coefficients used in the model.

Model testing is best done using a second set of forest data which was not used during the preparation of the yield model. The model is used to predict the behaviour of the forest from where the test data were collected and the results are compared with the actual observations. It is often necessary to repeat this stage several times. Adjustments to the model are made as a result of anomalies, or irregularities, that show up at each stage of testing.

Application of the Yield Model to the Required End-Use

A yield model may be applied in one of three ways:

- To forecast timber out-turn through a simple table or graph or a set of both. These can be used by forest planners directly or tables can be entered into a computer for updating of inventory data.
- To test forest management options as a computer or calculator programme which produces a table or graph of growth and yield for a particular set of treatments.
- To provide information on timber outputs for wider aspects of forest planning.

Examples of Yield Prediction Modelling Technology

A Simple Diameter Class Growth Projection Model.

Data Assembly

Assemble the relevant tree population and size data collected from a CFI for a whole forest management unit, or a part of it (a felling series), for which a yield determination is proposed.

Stand Table Preparation

Using PSP summary data, compile a tree population/diameter class distribution. DBH classes are typically in 5 cm intervals, e.g. 30-35 cm. Stratification of the tree population into smaller groupings can be made, based on species groups, e.g. "present commercial", "potentially commercial" and "presently non-commercial" species, and also on crown illumination classes.

Growth and Mortality Rates

Determine average diameter growth and average tree mortality rates from PSP data, for each diameter class.

Diameter Class Projection

Growth is projected for a five year period for each diameter class, on a per hectare basis, by applying growth and mortality rates. An adjustment is made to reflect the actual rather than the step-wise diameter class distribution of a stand because there are fewer trees in the upper part of a diameter class than in the lower part. In its simplest form, diameter class growth for each separate diameter class can be projected by applying the following formula:

$$N = S.I.\left(\frac{1}{q}\right)^{(1-I)}$$

where:

N = the number of trees growing from one diameter class to the next,

S = number of live trees in each diameter class,

I = average diameter increment for the class (cm),

q = an adjustment factor for each specific forest type and groups of diameter classes that is used to reflect the actual diameter class distribution.

The calculations can be carried out using standard spreadsheet software on a personal computer.

DIPSIM (Dipterocarp Forest Growth Simulation Model): an Empirical Individual-tree Growth Simulation Model.

DIPSIM is an empirical individual-tree growth simulation model that has been developed in

Sabah, Malaysia, as a management planning tool for natural Dipterocarp forests specifically for the purpose of:

- Predicting annual growth in terms of stocking, volume and basal area.
- Predicting changes in stand characteristics for periods of up to 60 years.
- Providing decision-support in yield regulation through the simulation of different harvesting prescriptions.

Main Features of the DIPSIM Modelling Process

Diameter Growth

Developing a multiple regression that relates basal area increment of individual trees to tree basal area, site quality, stand basal area and overtopping basal area. Increment equations have been developed for groups of species having similar growth patterns; 20 growth groups in all. Diameter increment patterns are developed for each species group.

Mortality

Developing a model that reflects regular and catastrophic mortality and predicts the probability of mortality from tree size and stand competition.

Recruitment

The model predicts recruitment rates for species in seven timber groups into the lowest (10 cm) stem diameter class.

Harvesting

Three steps are included in the harvesting component:

- Stocking assessment - the model compares the actual stocking of a stand against pre-defined post-harvest minimum stocking standards. In Sabah, a post-harvest minimum stocking of at least 8-10 trees/ha _ 60 cm dbh after 40 - 60 years.
- Tree selection and removal - predetermined minimum and maximum diameter cutting limits of 60 cm and 120 cm respectively are specified in the model.
- Harvesting damage - the user can specify in the model the percentage of trees expected to be destroyed in two size classes, 10-39 cm and_ 40 cm. The model randomly selects trees, removes them from the simulation database and places them into a mortality account database.

Model Operation

The PC-based DIPSIM is programmed in relational database software and the model structure consists of three main modules, database preparation, simulation and output. These are shown in Figure. The output module summarizes the simulation output in the following forms:

- Stand and stock tables.
- Volume increment tables.
- Harvesting tables, comprising a harvest composition table and an annual harvest record table.

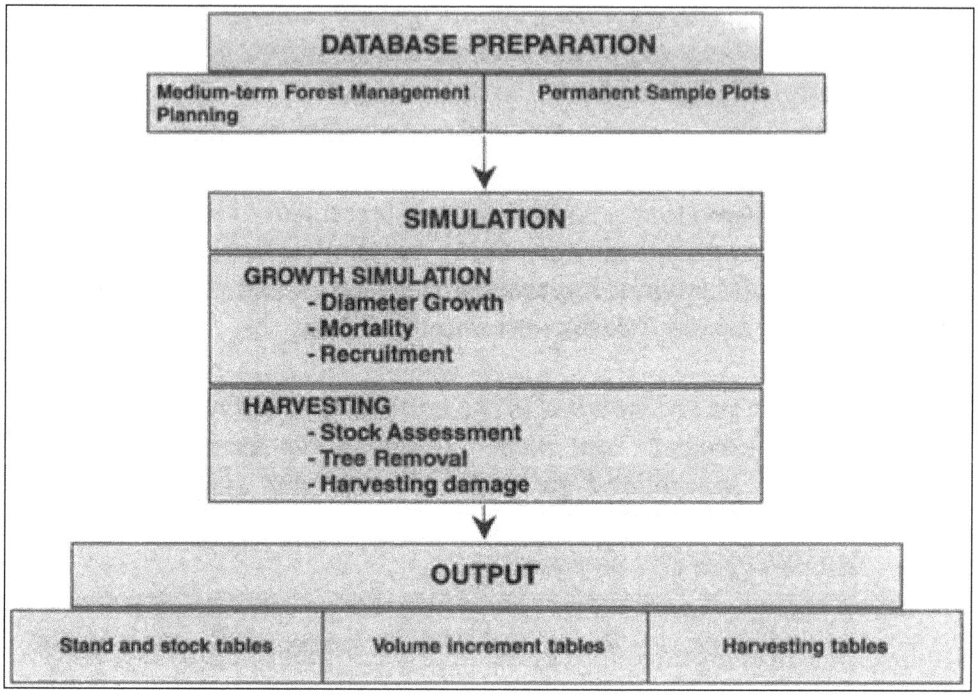

Fig: The DIPSIM programme structure.

Although practical and easy to understand, this and other diameter class projection methods do not effectively take changes in stand density or the consequences of spatial distribution (of diameter classes) of tropical forests into account. They are best used for preliminary yield analysis. For more accurate work the cohort modelling approach should be used.

Cohort Modelling

The term "cohort model" refers to a group of trees of the same species group and size class which is the basic information unit used for cohort growth modelling.

- CAFOGROM Model: A new and powerful tool for growth modelling research workers is the CAFOGROM model, developed at the Centro do Pesquisa Agropecuaria da Amazonia Oriental, in Brazil. It is a cohort model that is directed towards the analysis of PSP data from mixed tropical forests to produce growth and yield models. Although the CAFOGROM model is complex and does not provide, in itself, a standard set of computer programmes for the analysis of PSP data, the method is described in order that this new approach can become more widely known and to encourage local research on growth modelling. The four development stages of the CAFOGROM model are explained in figure.

Data Entry, Editing and Preliminary Processing:
Topics: Design of a PSP database, data checking, calculation of basic variables, such as increment, derivation of tree competition indices, formatting and condensing data for statistical analysis. Application: Establishment of a PSP data processing system, including all preliminary calculations for functional or transition matrix development.

Diameter Class Projection:
Topics: Formation of a stand table as a description of forest structure, mathematical and classical stand projection methods by diameter classes, matrix methods, pragmatic adaptations and case study for a complete model. Application: Description of analysis methods leading to either a spreadsheet stand projection model, or transition matrix model. Suitable for simple, highly aggregated models for broad forest-sector planning or feasibility studies.

Cohort Models for Forest Growth:
Topics: Analysis of growth functions for diameter increment, mortality and recruitment. Modelling harvesting operations and felling damage.
Application: Development of a dynamic simulation model for tropical mixed forests based on the cohorts/growth function approach. Suitable where good PSP data for many plots are available.

Growth Model Validation and Application:
Topics: Methods of validating functions by residual analysis. Use of growth models for forest planning, compatibility of inventory data, management information systems and the role of a growth model in them.
Application: Validation and application of a growth model as a component of a forest management system for use in the production of cutting plans in conjunction with inventory data.

Fig: Main Features of the CAFOGROM Modelling Process.

- SIRENA Model: In early 1996 a model called SIRENA I was developed for applied forest management in Costa Rica, Central America. It is an operational growth modelling tool, conceptually similar to CAFOGROM, but based on the analysis of local PSPs. SIRENA I is being used in the northern part of Costa Rica by an NGO CODEFORSA (Comisión de Desarrollo Forestal de San Carlos/Commission for Forest Development in San Carlos) for making management plans for private forest owners in that region of the country. SIRENA is based on "Excel" and "Visual Basic for Applications" software. SIRENA II is a revised version of SIRENA I based on operational experience with SIRENA I which is a flexible operational tool where a user may specify treatments and outputs. It is also used in Costa Rica.
- Queensland Model: The Queensland model, used until harvesting was discontinued for environmental conservation reasons in 1988, was developed to predict the growth of commercially loggable harvests in mixed species, uneven-aged rainforests in North Queensland, Australia. More than 100 commercial species and several hundred others are aggregated into about 20 species groups based on growth habit, volume relations and commercial criteria. Trees are grouped into cohorts, according to species group and tree size, which form the basis for simulation. Equations have been developed to predict increment, mortality and recruitment. The model has been used for determining wood harvests from north Queensland forests and provide an objective basis for investigating the effects of rain forest management strategies. The approach is expected to be applicable to other natural tropical forests.

References

- Forestry, explore: edugreen.teri.res.in, Retrieved 19, March 2020
- 12-reasons-why-forests-are-important: glo-be.be, Retrieved 10, June 2020
- What-are-forests-explain-types-of-forest: einsty.com, Retrieved 11, April 2020
- Forest, environment, forest-conservation-useful-methods-for-forest-conservation-25277: yourarticlelibrary.com, Retrieved 16, March 2020

Chapter 3

Afforestation

Afforestation can be defined as the act or process of planting trees with the ultimate aim of establishing a forest on land not previously forested. Reforestation refers to the process of replanting trees in forested and woodland areas which have undergone depletion. Urban forestry, biosequestration and climate change mitigation are other topics related to afforestation which are discussed at length in this chapter.

An afforestation project in Rand Wood, Lincolnshire, England.

Afforestation is the establishment of a forest or stand of trees in an area where there was no previous tree cover. Reforestation is the reestablishment of forest cover, either naturally (by natural seeding, coppice, or root suckers) or artificially (by direct seeding or planting).

Forestation is the establishment of forest growth on areas that either had forest or lacked it. Reforestation and afforestation are categories of forestation. Many governments and non-governmental organizations directly engage in programs of *afforestation* to create forests, increase carbon capture and carbon sequestration, and help to anthropogenically improve biodiversity. (In the UK, afforestation may mean converting the legal status of some land to "royal forest".) Special tools, e.g. tree planting bar, are used to make planting of trees easier and faster.

Biological Process

Gap dynamics is the pattern of plant growth that occurs following the creation of a forest gap, a local area of natural disturbance that results in an opening in the canopy of a forest. Gap dynamics are a typical characteristic of temperate and tropical forests, and have a wide variety of causes and effects on forest life.

In Areas of Degraded Soil

In some places, forests need help to reestablish themselves because of environmental factors. For example, in arid zones, once forest cover is destroyed, the land may dry and become inhospitable to new tree growth. Other factors include overgrazing by livestock, especially animals such as goats, cows, and over-harvesting of forest resources. Together these may lead to desertification and the loss of topsoil; without soil, forests cannot grow until the long process of soil creation has been completed - if erosion allows this. In some tropical areas, forest cover removal may result in a duricrust or duripan that effectively seal off the soil to water penetration and root growth. In many areas, reforestation is impossible because people are using the land. In other areas, mechanical breaking up of duripans or duricrusts is necessary, careful and continued watering may be essential, and special protection, such as fencing, may be needed.

Countries and Regions

Afforested botanical garden in Hattori Ryokuchi Park, Japan.

Brazil

There is extensive and ongoing Amazon deforestation.

China

China has deforested most of its historically wooded areas. China reached the point where timber yields declined far below historic levels, due to over-harvesting of trees beyond sustainable yield. Although it has set official goals for reforestation, these goals are set over an 80-year time horizon and have not been significantly met by 2008. China is trying to correct these problems by projects as the Green Wall of China, which aims to replant a great deal of forests and halt the expansion of the Gobi desert. A law promulgated in 1981 requires that every school student over the age of 11 plants at least one tree per year. As a result, China has the highest afforestation rate of any country or region in the world, with 47,000 square kilometers of afforestation in 2008. However, the forest area per capita is still far lower than the international average. There has also been considerable criticism regarding the effectiveness of planting so many trees especially in regions where they never grew prior. Studies reveal that the water table of those areas is becoming deeper indicating significant water loss.

India

Afforestation in South India

India has witnessed a minor increase in the percentage of the land area under forest cover from 1950 to 2006. In 1950 around 40.48 million hectares was covered by forest. In 1980 it increased to 67.47 million hectares and in 2006 it was found to be 69 million hectares. 23% of India is covered by forest. The forests of India are grouped into 5 major categories and 16 types based on biophysical criteria. 38% of forest is categorised as subtropical dry deciduous and 30% as tropical moist deciduous plus other smaller groups. It is taken care that only local species are planted in an area. Trees bearing fruits are preferred wherever possible due to their function as a food source.

Hong Kong

Since the founding of the crown colony in the 19th century, afforestation has taken place to prevent soil erosion in the catchment areas of the reservoirs that were built. During the Japanese occupation in the Second World War, the countryside was deforested as the remaining population required fuel to survive. Most of the trees were cut down and extensive reafforestation was carried out after the war. Trees that were planted are mostly non-native species, such as: Pinus massoniana, Acacia confusa (Formosan acacia), Lophostemon confertus and the Paper Bark Tree.

Burkina Faso

Desertification is increasing along the Sahel, the strip of land between Africa's fertile tropics and the Sahara Desert. After a crippling famine in the 1970s caused by overgrazing and deforestation, a local community approach has been pioneered by Yacouba Sawadogo, a peasant farmer. By replanting trees and crops together in holes filled with compost, whole villages have been able to move back to areas considered uninhabitable.

Iran

Iran is considered a low forest cover region of the world with present cover approximating seven percent of the land area. This is a value reduced by an estimated six million hectares of virgin forest, which includes oak, almond and pistachio. Due to soil substrates, it is difficult to achieve afforestation on a large scale compared to other temperate areas endowed with more fertile and less rocky and arid soil conditions. Consequently, most of the afforestation is conducted with non-native species, leading to habitat destruction for native flora and fauna, and resulting in an accelerated loss of biodiversity.

JNF trees in the Negev Desert. Man-made dunes (here a liman) help keep in rainwater, creating an oasis.

Israel

Tree-planting is an ancient Jewish tradition, mentioned in the Talmud as being more important than greeting the Messiah. With over 240 million planted trees, Israel is one of only two countries that entered the 21st century with a net gain in the number of trees, due to massive afforestation efforts. Israeli forests are the product of a major afforestation campaign by the Jewish National Fund (JNF).

Critics argue that many JNF lands inside the West Bank were illegally confiscated from Palestinian refugees, and that the JNF furthermore should not be involved with lands in the West Bank. Shaul Ephraim Cohen has claimed that trees have been planted to restrict Bedouin herding. Susan Nathan wrote that forests were planted on the site of abandoned Arab villages after the 1948 war.

Since 2009, the JNF has provided the Palestinian Authority with 3,000 tree seedlings for a forested area being developed on the edge of the new city of Rawabi, north of Ramallah.

North Africa

In North Africa, the Sahara Forest Project coupled with the Seawater greenhouse has been proposed. Some projects have also been launched in countries as Senegal to revert desertification. As of 2010, African leaders are discussing the combining of national resources to increase effectiveness. In addition, other projects as the Keita Project in Niger have been launched in the past, and have been able to locally revert damage done by desertification.

Europe

Europe has deforested the majority of its historical forests. The European Union (EU) has paid farmers for afforestation since 1990, offering grants to turn farmland back into forest and payments for the management of forest. Between 1993 and 1997, EU afforestation policies made possible the re-forestation of over 5,000 square kilometres of land. A second program, running between 2000 and 2006, afforested more than 1000 square kilometres of land (precise statistics not yet available). A third such program began in 2007. Europe's forests are growing by 0.8 million ha a year thanks to these programmes.

In Poland, the National Program of Afforestation was introduced by the government after World War II, when area of forests shrank to 20% of country's territory. Consequently, forested areas of Poland grew year by year, and on December 31, 2006, forests covered 29% of the country. It is planned that by 2050, forests will cover 33% of Poland.

According to Food and Agriculture Organization statistics, Spain had the third fastest afforestation rate in Europe in the 1990-2005 period, after Iceland and Ireland. In those years, a total of 44,360 square kilometers were afforested, and the total forest cover rose from 13,5 to 17,9 million hectares. In 1990, forests covered 26.6% of the Spanish territory. As of 2007, that figure had risen to 36.6%. Spain today has the fifth largest forest area in the European Union.

In January 2013 the UK government set a target of 12% woodland cover in England by 2060, up from the then 10%. Government-backed initiatives such as the Woodland Carbon Code are intended to support this objective by encouraging corporations and landowners to create new woodland to offset their carbon emissions.

Alpine and Subalpine regions have undergone a lot of deforestation and then forestation in the last 300 years. Out of this has emerged much practical experience. One example is the Rotten group, which is a method to bring in stable age mixed tree communities.

Australia

In Adelaide, South Australia (a city of 1.3 million), Premier Mike Rann (2002 to 2011) launched an urban forest initiative in 2003 to plant 3 million native trees and shrubs by 2014 on 300 project sites across the metro area. The projects range from large habitat restoration projects to local biodiversity projects. Thousands of Adelaide citizens have participated in community planting days. Sites include parks, reserves, transport corridors, schools, water courses and coastline. Only trees native to the local area are planted to ensure genetic integrity. Premier Rann said the project aimed to beautify and cool the city and make it more liveable; improve air and water quality and reduce Adelaide's greenhouse gas emissions by 600,000 tonnes of Co2 a year. He said it was also about creating and conserving habitat for wildlife and preventing species loss.

United States

The United States is roughly one-third covered in forest and woodland. Nevertheless, areas in the US were subject to significant tree planting. In the 1800s people moving westward encountered the Great Plains – land with fertile soil, a growing population and a demand for timber but with few trees to supply it. So tree planting was encouraged along homesteads. Arbor Day was founded in 1872 by Julius Sterling Morton in Nebraska City, Nebraska. By the 1930s the Dust Bowl environmental disaster signified a reason for significant new tree cover. Public works programs under the New Deal saw the planting of 18,000 miles of windbreaks stretching from North Dakota to Texas to fight soil erosion.

At their summit in Copenhagen in 2009, organised by the UK based The Climate Group, leaders of subnational governments – states, regions and provinces – unanimously supported a recommendation by Premier Rann to plant one billion trees across their varied jurisdictions. The initiative was strongly supported by leaders present including Quebec Premier Jean Charest, California Governor Arnold Schwarzenegger and Scottish First Minister Alex Salmond. At a subsequent meeting in Rio de Janeiro in June 2012, The Climate Group announced that it had already received commitments by member governments to plant more than 500 million trees.

Gap Dynamics

Treefall gaps in the Amazon allow sunlight to reach the forest floor.

Gap dynamics refers to the pattern of plant growth that occurs following the creation of a forest gap, a local area of natural disturbance that results in an opening in the canopy of a forest. Gap dynamics are a typical characteristic of both temperate and tropical forests and have a wide variety of causes and effects on forest life.

Gaps are the result of natural disturbances in forests, ranging from a large branch breaking off and dropping from a tree, to a tree dying then falling over, bringing its roots to the surface of the ground, to landslides bringing down large groups of trees. Because of the range of causes, gaps, therefore, have a wide range of sizes, including small and large gaps. Regardless of size, gaps allow an increase in light as well as changes in moisture and wind levels, leading to differences in microclimate conditions compared to those from below the closed canopy, which are generally cooler and more shaded.

For gap dynamics to occur in naturally disturbed areas, either primary or secondary succession must occur. Ecological secondary succession is much more common and pertains to the process of vegetation replacement after a natural disturbance. Secondary succession results in second-growth or secondary forest, which currently covers more of the tropics than old-growth forest.

Since gaps let in more light and create diverse microclimates, they provide the ideal location and conditions for rapid plant reproduction and growth. In fact, most plant species in the tropics are dependent, at least in part, on gaps to complete their life cycles.

Disturbances

Gap dynamics are the result of disturbances within an ecosystem. There are both large scale and small scale disturbances, and both are influenced by duration and frequency. These all affect the resulting impact and regeneration patterns of the ecosystem.

Broken trees create gaps in the central Amazon.

The most common type of disturbance within a tropical ecosystem is fire. Since most nutrients in a tropical ecosystem are contained in the biomass of plants, fire is an important component of recycling these nutrients and therefore regenerating an ecosystem.

An example of a small scale disturbance is a tree falling. This can cause soil movement, which re-distributes any nutrients or organisms that were attached to the tree. The tree falling also opens up the canopy for light entrance, which can support the growth of other trees and plants.

After a disturbance, there are several ways in which regeneration can occur. One way, termed the advance regeneration pathway, is when the primary understory already contains seedlings and saplings. This method is most common in the Neotropics when faced when small scale disturbances. The next pathway is from tree remains, or any growth from bases or roots, and is common in small disturbance gaps. The third route is referred to as the soil seed bank, and is the result of germination of seeds already found in the soil. The final regeneration pathway is the arrival of new seeds via animal dispersal or wind movement. The most critical components of the regeneration are seed distribution, germination, and survival.

Forest Gaps and Forest Regeneration

Until recently, forest regeneration practices in North America have largely followed an agricultural model, with research concentrated on techniques for establishing and promoting early growth of planted stock after clearcutting, followed by studies of growth and yield emphasizing single-species growth uninfluenced by overstorey canopy. Coates (2000) questioned this approach and proposed a shift to a more ecologically and socially based approach able to accommodate greater diversity in managed stands. Predictive models of forest regeneration and growth that take account of variable levels of canopy retention will be needed as the complexity of managed forest stands increases.

Tree regeneration occurring inside canopy gaps after disturbance has been studied widely. Studies of gap dynamics have contributed much to an understanding of the role of small-scale disturbance in forest ecosystems, but they have been little used by foresters to predict tree responses following partial cutting.

In high-latitude northern forests, position inside a gap can have a pronounced effect on resource levels (e.g., light availability) and microclimate conditions (e.g., soil temperature), especially along the north–south axis. Such variation must inevitably affect the amount and growth of regeneration; but relying solely on natural regeneration to separate the effects of gap size and position is problematic (Coates 2000). Among the many factors affecting seedling establishment following canopy disturbance are parent tree proximity and abundance, seedbed substrate, presence of seed consumers and dispersers, and climatic and microclimatic variability. Planted trees can be used to avoid many of the stochastic events surrounding natural seedling establishment.

Gradients of canopy influence can be created by partial cutting, and tree growth responses within gaps of various sizes and configurations, as well as within the adjacent forest matrix can form a basis for tree species selection. Hybrid spruce (the complex of white spruce, Sitka spruce, and occasionally Engelmann spruce) was one of several coniferous species used in a study in the Moist Cold subzone of the Interior Cedar–Hemlock zone in northwestern British Columbia. A total of 109 gaps were selected from a population of openings created by logging within each light and heavy partial cutting treatment in stands averaging 30 m in canopy height; 76 gaps were less than 1000 m^2, 33 were between 1000 m^2 and 5000 m^2. Canopy gap size was calculated as the area of an ellipse, the major axis of which was the longest line that could be run from canopy edge to canopy edge inside the gap, and the minor axis was the longest line that could be run from canopy edge perpendicular to the long line. Seedlings were planted in gaps and in the undisturbed and clearcut treatment units. There were strong and consistent trends in growth response among the seedlings as gap size increased. In all species, growth increased rapidly from small single-tree gaps to about 1000 m^2, but thereafter, there was little change up to 5000 m^2. Tree size and current growth rates for all species were highest in full open conditions. In large and medium gaps (300–1000 m^2), the largest trees of all species occurred in the middle gap position, with little difference between the sunny north and shady south positions, lodgepole pine excepted. The light advantage expected off the north end of higher-latitude gaps was not a benefit for tree growth, suggesting that below-ground effects of canopy edge trees have an important influence of seedling growth in these forests.

In a study near Chapleau, Ontario (Groot 1992, Groot et al. 1997), openings were created in 40-year-old aspen and monitored to determine their influence on outplanted white spruce seedling development. Circular openings 9 m and 18 m in diameter, 9 m and 18 m wide east–west strips, and a 100 m × 150 m clearcut were planted and spot-seeded. The variation in solar radiation, air temperature, and soil temperature among the strips and plots was almost as great as the variation between the clearcut and intact forest. Solar radiation during the first growing season varied from 18% of the above-canopy values within the uncut stand to 68% values at the center of the 18 m strip. Near the edges of the strips, solar radiation was about 40% of the above-canopy along the south and 70% to 80% along the north. Stomatal conductance in white spruce seedlings declined generally from more sheltered to more exposed environments, correlating best with increased vapor pressure deficit (VPD). Without vegetation control, position in openings had little effect on the growth of planted white spruce; regrowth of lesser vegetation isolated seedlings from the microclimatic effects of overstorey treatment. Seedling diameters were independent of environment, while height growth was only slightly greater in environments having more light. With vegetation control, white spruce diameter and height were greatest in the center of the strips, even though there was less light there than along the north edge of the strips. Moisture stress may have accounted for that result.

Primary Succession

Succession is the slow rebuilding of forest gaps from natural or human disturbances. When major geological changes such as volcano eruptions or landslides occur, the current vegetation and soil may erode away leaving only rock. Primary succession occurs when pioneer species such as lichens colonize rock. As the lichens and mosses decompose, a soil substrate forms called peat. The peat, over time, will create a terrestrial ecosystem. From there on herbaceous, non-woody plants will develop and trees will follow. Major holes or gaps in the forest ecosystem will take hundreds of years to regenerate from a rock base.

Secondary Succession

Cecropia trees are a common pioneer species found in gaps.

Secondary succession occurs where a disturbance has taken place but soil remains and is able to support plant growth. It does not take nearly as long for plant regeneration to occur because of the soil substrate already present. Secondary succession is much more common than primary succession in the tropics.

Ecological secondary succession occurs in four distinct phases: First, rapid colonization of cleared land by species such as herbs, shrubs, and climbers as well as seedlings from pioneer tree species occurs and this can last up to three years. After that, short lived but fast growing shade intolerant species form a canopy over 10 to 30 years. Non-pioneer heliophilic (sun-loving) tree species then add to the biomass and species richness as well as shade tolerant species and this can last 75 to 150 years. Finally, shade-tolerant species regain full canopy stature indefinitely until another major disturbance occurs.

Secondary succession in the tropics begins with pioneer species, which are rapidly growing and include vines and shrubs. Once these species are established, large heliophilic species will develop such as heliconias. Cecropias are also a major pioneering tree in the tropics and they are adapted to grow well where forest gaps are giving way to sunlight. Shade-tolerant species that have remained low in the forest develop and become much taller. These successional phases do not have definite order or structure and because of the very high biodiversity in the tropics, there is a lot of competition for resources such as soil nutrients and sunlight.

Examples of Tree Dynamics

Due to the fact that horizontal and vertical heterogeneity of a forest is significantly increased by gaps, gaps become an obvious consideration in explaining high biodiversity. It has been proven that gaps create suitable conditions for rapid growth and reproduction. For example, non-shade tolerant plant species and many shade-tolerant plant species respond to gaps with an increase in growth, and at least a few species are dependent on gaps to succeed in their respected environments (Brokaw 1985; Hubbell and Foster 1986b; Murray 1988; Clark and Clark 1992). Gaps create diverse microclimates, affecting light, moisture, and wind conditions (Brokaw 1985). For example, exposure to edge effects increases a microclimate's light and wind intensity and decreases its moisture. A study conducted on Barro Colorado Island in Panama showed that gaps had greater seedling establishment and higher sapling densities than control areas.

Species richness was higher in gaps than in control areas, and there was more diversity in species composition among gaps. However, this study also found that there was a low recruitment rate per gap, which explains why gaps differed in species composition. With 2% to 3% for pioneer species and 3% to 6% for shade-tolerant and intermediate species. Suggesting that most species could not take advantage of gaps because they couldn't get to them through seed dispersal. With that said, the Janzen-Connell effect plays a major role in the tree species' relationship with gaps. The Janzen-Connell density dependent mortality model states that most trees die as seed or seedlings. In addition, host-specific predators or pathogens are predicted to be greatest where density is greatest, which is underneath parent tree. This corroborates with the major causes of gaps, which are the falling of trees due to mortality caused by termites or epiphyte growth. The Janzen-Connell model also states that balance between dispersal distance and mortality should cause highest recruitment to be at a certain distance away from the parent. Therefore, if these gaps are being created by the parents, the seedlings recruit away from the gap, resulting in increasing survival rates as the distance from the parent increases. This explains the low recruitment rate per gap found in the experiment conducted in Barro Colorado Island.

In corroboration, a study conducted in La Selva in Costa Rica calculated the crown illumination index for nine tree species ranging from gap specialists to emergent canopy species. Crown illumination values ranged from 1, which indicated low light, and 6, which indicated that the tree crown was completely exposed . After using a mathematical model to calculate the changes in tree diameter and changes in crown illumination with age. This model helped estimate life expectancy, time of passage to various sizes, and age patterns of mortality. The results showed what most gap dynamics studies show, pioneer species thrived in high light environments and non-pioneer species showed high mortality when young but the rate of mortality decreased as they aged. However, once trees were very large survivorship then decreased.

Disturbance (Ecology)

In biology, a disturbance is a temporary change in environmental conditions that causes a pronounced change in an ecosystem. Disturbances often act quickly and with great effect, to alter the physical structure or arrangement of biotic and abiotic elements. Disturbance can also occur over a long period of time and can impact the diversity within an ecosystem. Major ecological disturbances may include fires, flooding, windstorms, insect outbreaks and trampling. Earthquakes, various types of volcanic eruptions, tsunami, firestorms, impact events, climate change, and the devastating effects of human impact on the environment (anthropogenic disturbances) such as clearcutting, forest clearing and the introduction of invasive species can be considered major disturbances. Disturbance forces can have profound immediate effects on ecosystems and can, accordingly, greatly alter the natural community. Because of these and the impacts on populations, disturbance determines the future shifts in dominance, various species successively becoming dominant as their life history characteristics, and associated life-forms, are exhibited over time.

The disturbance of a fire can clearly be seen by comparing the unburnt (left) and burnt (right) sides of the mountain range in South Africa. The veld ecosystem relies on periodic fire disturbances like these to rejuvenate its self.

Criteria

Conditions under which natural disturbances occur are influenced mainly by climate, weather, and location. Fire disturbances will only occur in areas where there is low precipitation, some form of ignition (typically lightning), and enough flammable biomass to allow fire to spread. Conditions often occur as part of a cycle and disturbances may be periodic. Other disturbances, such as those caused by humans, invasive species or impact events, can occur anywhere and are not necessarily cyclic. Extinction vortices may result in multiple disturbances or a greater frequency of a single disturbance. Immediately after a disturbance there is a pulse of recruitment or regrowth under conditions of little competition for space or other resources. After the initial pulse, recruitment slows since once an individual plant is established it is very difficult to displace. Due to the varying forms of disturbance this directly impacts the organisms which will exploit the disturbance and create diversity within an ecosystem.

Cyclic disturbance

Often, when disturbances occur naturally, they provide conditions that favor the success of different species over pre-disturbance organisms. This can be attributed to physical changes in the biotic and abiotic conditions of an ecosystem. Because of this, a disturbance force can change an ecosystem for significantly longer than the period over which the immediate effects persist. With the passage of time following a disturbance, shifts in dominance may occur with ephemeral herbaceous life-forms progressively becoming over topped by taller perennials herbs, shrubs and trees. However, in the absence of further disturbance forces, many ecosystems trend back toward pre-disturbance conditions. Long lived species and those that can regenerate in the presence of their own adults finally become dominant. Such alteration, accompanied by changes in the abundance of different species over time, is called ecological succession. Succession often leads to conditions that will once again predispose an ecosystem to disturbance.

Damages of storm Kyrill in Wittgenstein, Germany.

Pine forests in the western North America provide a good example of such a cycle involving insect outbreaks. The mountain pine beetle (*Dendroctonus ponderosae*) play an important role in limiting pine trees like lodgepole pine in forests of western North America. In 2004 the beetles affected more than 90,000 square kilometres. The beetles exist in endemic and epidemic phases. During epidemic phases swarms of beetles kill large numbers of old pines. This mortality creates openings in the forest for new vegetation. Spruce, fir, and younger pines, which are unaffected by the beetles, thrive in canopy openings. Eventually pines grow into the canopy and replace those lost. Younger pines are often able to ward off beetle attacks but, as they grow older, pines become less vigorous and more susceptible to infestation. This cycle of death and re-growth creates a temporal mosaic of pines in the forest. Similar cycles occur in association with other disturbances such as fire and windstorms.

When multiple disturbance events affect the same location in quick succession, this often results in a "compound disturbance," an event which, due to the combination of forces, creates a new situation which is more than the sum of its parts. For example, windstorms followed by fire can create fire temperatures and durations that are not expected in even severe wildfires, and may have surprising effects on post-fire succession. Environmental stresses can be described as pressure on the environment, with compounding variables such as extreme temperature or precipitation changes—which all play a role in the diversity and succession of an ecosystem. With environmental moderation, diversity increases because of the intermediate- disturbance effect, decreases because of the competitive-exclusion effect, increases because of the prevention of competitive exclusion by moderate predation, and decreases because of the local extinction of prey by severe predation. A reduction in recruitment density reduces the importance of competition for a given level of environmental stress.

Species Adapted to Disturbance

A disturbance may change a forest significantly. Afterwards, the forest floor is often littered with dead material. This decaying matter and abundant sunlight promote an abundance of new growth. In the case of forest fires a portion of the nutrients previously held in plant biomass is returned quickly to the soil as biomass burns. Many plants and animals benefit from disturbance conditions. Some species are particularly suited for exploiting recently disturbed sites. Vegetation with the potential for rapid growth can quickly take advantage of the lack of competition. In the northeastern United States, shade-intolerant trees like pin cherry and aspen quickly fill in forest gaps created by fire or windstorm (or human disturbance). Silver maple and eastern sycamore are similarly well adapted to floodplains. They are highly tolerant of standing water and will frequently dominate floodplains where other species are periodically wiped out.

Forest fire burns on the island of Zakynthos in Greece on July 25th, 2007.

When a tree is blown over, gaps typically are filled with small herbaceous seedlings but, this is not always the case; shoots from the fallen tree can develop and take over the gap. The sprouting ability can have major impacts on the plant population, plant populations that typically would have exploited the tree fall gap get over run and can not complete against the shoots of the fallen tree. Species adaptation to disturbances is species specific but how each organism adapts effects all the species around them.

Another species well adapted to a particular disturbance is the Jack Pine in boreal forests exposed to crown fires. They, as well as some other pine species, have specialized serotinous cones that only open and disperse seeds with sufficient heat generated by fire. As a result, this species often dominates in areas where competition has been reduced by fire.

Species that are well adapted for exploiting disturbance sites are referred to as pioneers or early successional species. These shade-intolerant species are able to photosynthesize at high rates and as a result grow quickly. Their fast growth is usually balanced by short life spans. Furthermore, although these species often dominate immediately following a disturbance, they are unable to compete with shade-tolerant species later on and replaced by these species through succession. However these shifts may not reflect the progressive entry to the community of the taller long-lived forms, but instead, the gradual emergence and dominance of species that may have been present, but inconspicuous directly after the disturbance.

While plants must deal directly with disturbances, many animals are not as immediately affected by them. Most can successfully evade fires, and many thrive afterwards on abundant new growth on the forest floor. New conditions support a wider variety of plants, often rich in nutrients compared to pre-disturbance vegetation. The plants in turn support a variety of wildlife, temporarily increasing biological diversity in the forest.

Importance

Biological diversity is dependent on natural disturbance. The success of a wide range of species from all taxonomic groups is closely tied to natural disturbance events such as fire, flooding, and windstorm. As an example, many shade-intolerant plant species rely on disturbances for successful establishment and to limit competition. Without this perpetual thinning, diversity of forest flora can decline, affecting animals dependent on those plants as well.

A good example of this role of disturbance is in ponderosa pine (*Pinus ponderosa*) forests in the western United States, where surface fires frequently thin existing vegetation allowing for new growth. If fire is suppressed, douglas fir (*Pesudotsuga menziesii*), a shade tolerant species, eventually replaces the pines. Douglas firs, having dense crowns, severely limit the amount of sunlight reaching the forest floor. Without sufficient light new growth is severely limited. As the diversity of surface plants decreases, animal species that rely on them diminish as well. Fire, in this case, is important not only to the species directly affected but also to many other organisms whose survival depends on those key plants.

Diversity is low in harsh environments because of the intolerance of all but opportunistic and highly resistant species to such conditions. The interplay between disturbance and these biological processes seems to account for a major portion of the organization and spatial patterning of natural communities. Disturbance variability and species diversity are heavily linked, and as a result require adaptations that help increase plant fitness necessary for survival.

Reforestation

Reforestation is the natural or intentional restocking of existing forests and woodlands that have been depleted, usually through deforestation. Reforestation can be used to rectify or improve the quality of human life by soaking up pollution and dust from the air, rebuild natural habitats and ecosystems, mitigate global warming since forests facilitate biosequestration of atmospheric carbon dioxide, and harvest for resources, particularly timber, but also non-timber forest products.

Tropical tree nursery at Planeta Verde Reforestación S.A.'s plantation in Vichada Department, Colombia.

A 21-year-old plantation of red pine in Southern Ontario.

The term reforestation is similar to afforestation, the process of restoring and recreating areas of woodlands or forests that may have existed long ago but were deforested or otherwise removed at some point in the past. Sometimes the term *re-afforestation* is used to distinguish between the original forest cover and the later re-growth of forest to an area. Forestation is the establishment of forest growth on areas that either had forest or lacked it.

Special tools, e.g. tree planting bars, are used to make planting of trees easier and faster.

Management

A debated issue in managed reforestation is whether or not the succeeding forest will have the same biodiversity as the original forest. If the forest is replaced with only one species of tree and all other vegetation is prevented from growing back, a monoculture forest similar to agricultural crops would be the result. However, most reforestation involves the planting of different seedlots of seedlings taken from the area, often of multiple species. Another important factor is the natural regeneration of a wide variety of plant and animal species that can occur on a clear cut. In some areas the suppression of forest fires for hundreds of years has resulted in large single aged and single species forest stands. The logging of small clear cuts and/or prescribed burning, actually increases the biodiversity in these areas by creating a greater variety of tree stand ages and species.

For Harvesting

Reforestation need not be only used for recovery of accidentally destroyed forests. In some countries, such as Finland, many of the forests are *managed* by the wood products and pulp and paper industry. In such an arrangement, like other crops, trees are planted to replace those that have been cut. In such circumstances, the industry can cut the trees in a way to allow easier reforestation. The wood products industry systematically replaces many of the trees it cuts, employing large numbers of summer workers for tree planting work. For example, in 2010, Weyerhaeuser reported planting 50 million seedlings. However replanting an old-growth forest with a plantation is not replacing the old with the same characteristics in the new.

In just 20 years, a teak plantation in Costa Rica can produce up to about 400 m^3 of wood per hectare. As the natural teak forests of Asia become more scarce or difficult to obtain, the prices commanded by plantation-grown teak grows higher every year. Other species such as mahogany grow more slowly than teak in Tropical America but are also extremely valuable. Faster growers include pine, eucalyptus, and *Gmelina.*

Reforestation, if several indigenous species are used, can provide other benefits in addition to financial returns, including restoration of the soil, rejuvenation of local flora and fauna, and the capturing and sequestering of 38 tons of carbon dioxide per hectare per year.

The reestablishment of forests is not just simple tree planting. Forests are made up of a community of species and they build dead organic matter into soils over time. A major tree-planting program could enhance the local climate and reduce the demands of burning large amounts of fossil fuels for cooling in the summer.

For Climate Change Mitigation

Forests are an important part of the global carbon cycle because trees and plants absorb carbon dioxide through photosynthesis. By removing this greenhouse gas from the air, forests function as terrestrial carbon sinks, meaning they store large amounts of carbon. At any time, forests account for as much as double the amount of carbon in the atmosphere. Even as more anthropogenic carbon is produced, forests remove around three billion tons of anthropogenic carbon every year. This amounts to about 30% of all carbon dioxide emissions from fossil fuels. Therefore, an increase in the overall forest cover around the world would tend to mitigate global warming.

There are four major strategies available to mitigate carbon emissions through forestry activities: increase the amount of forested land through a reforestation process; increase the carbon density of existing forests at a stand and landscape scale; expand the use of forest products that will sustainably replace fossil-fuel emissions; and reduce carbon emissions that are caused from deforestation and degradation.

Achieving the first strategy would require enormous and wide-ranging efforts. However, there are many organizations around the world that encourage tree-planting as a way to offset carbon emissions for the express purpose of fighting climate change. For example, in China, the Jane Goodall Institute, through their Shanghai Roots & Shoots division, launched the Million Tree Project in Kulun Qi, Inner Mongolia to plant one million trees to stop desertification and help curb climate change. China has used 24 billion metres squared of new forest plantation and natural forest regrowth to offset 21% of Chinese fossil fuel emissions in 2000. In Java, Indonesia each newlywed couple is to give whoever is sermonizing their wedding 5 seedlings to combat global warming. Each couple that wishes to have a divorce has to give 25 seedlings to whoever divorces them.

The second strategy has to do with selecting species for tree-planting. In theory, planting any kind of tree to produce more forest cover would absorb more carbon dioxide from the atmosphere. On the other hand, a genetically modified tree specimen might grow much faster than any other regular tree. Some of these trees are already being developed in the lumber and biofuel industries. These fast-growing trees would not only be planted for those industries but they can also be planted to help absorb carbon dioxide faster than slow-growing trees.

Extensive forest resources placed anywhere in the world will not always have the same impact. For example, large reforestation programs in boreal or subarctic regions have a limited impact on climate mitigation. This is because it substitutes a bright snow-dominated region that reflects the sunlight with dark forest canopies. A study from the National Center for Atmospheric Research in Boulder, Colorado, USA, found that trees in temperate latitudes have a net warming effect on the atmosphere. The heat that dark leaves release without absorbing outweighs the carbon they sequester. On the other hand, a positive example would be reforestation projects in tropical regions, which would lead to a positive biophysical change such as the formation of clouds. These clouds would then reflect the sunlight, creating a positive impact on climate mitigation.

There is an advantage to planting trees in tropical climates with wet seasons. In such a setting, trees have a quicker growth rate because they can grow year-round. Trees in tropical climates have, on average, larger, brighter, and more abundant leaves than non-tropical climates. A study of the girth of 70,000 trees across Africa has shown that tropical forests are soaking up more carbon dioxide pollution than previously realized. The research suggests almost one fifth of fossil fuel emissions are absorbed by forests across Africa, the Amazon basin and Asia. Simon Lewis, a climate expert at the University of Leeds, who led the study, said: "Tropical forest trees are absorbing about 18% of the carbon dioxide added to the atmosphere each year from burning fossil fuels, substantially buffering the rate of change."

It is also important to deal with the rate of deforestation. At this point, there are 13 billion metres squared of tropical regions that are deforested every year. There is potential for these regions to reduce rates of deforestation by 50% by 2050, which would be a huge contribution to stabilize the global climate.

On Abandoned Farmland

With increased intensive agriculture and urbanization, there is an increase in the amount of abandoned farmland. By some estimates, for every half a hectare of original old-growth forest cut down, more than 20 hectares of new secondary forests are growing, even though they do not have the same biodiversity as the original forests and original forests store 60% more carbon than these new secondary forests. According to a study in Science, promoting regrowth on abandoned farmland could offset years of carbon emissions.

Promotion Strategies

Land Rights

Transferring land rights to indigenous inhabitants is argued to efficiently conserve forests.

Transferring rights over land from public domain to its indigenous inhabitants is argued to be a cost effective strategy to conserve forests. This includes the protection of such rights entitled in existing laws, such as India's Forest Rights Act. The transferring of such rights in China, perhaps the largest land reform in modern times, has been argued to have increased forest cover. In Brazil, forested areas given tenure to indigenous groups have even lower rates of clearing than national parks.

Incentives

Some incentives for reforestation can be as simple as a financial compensation. Streck and Scholz (2006) explain how a group of scientists from various institutions have developed a compensated reduction of deforestation approach which would reward developing countries that disrupt any further act of deforestation. Countries that participate and take the option to reduce their emissions from deforestation during a committed period of time would receive financial compensation for the carbon dioxide emissions that they avoided. To raise the payments, the host country would issue government bonds or negotiate some kind of loan with a financial institution that would want to take part in the compensation promised to the other country. The funds received by the country could be invested to help find alternatives to the extensive cutdown of forests. This whole process of cutting emissions would be voluntary, but once the country has agreed to lower their emissions they would be obligated to reduce their emissions. However, if a country was not able to meet their obligation, their target would get added to their next commitment period. The authors of these proposals see this as a solely government-to-government agreement; private entities would not participate in the compensation trades.

Examples

Forest regrowth in Mount Baker-Snoqualmie National Forest, Washington state, USA.

Sub-Saharan Africa

Plans to plant a nine-mile width of trees on the Southern Border of the Sahara Desert. The Great Green Wall initiative is a pan-African proposal to "green" the continent from west to east in order to battle desertification. It aims at tackling poverty and the degradation of soils in the Sahel-Saharan region, focusing on a strip of land of 15 km (9 mi) wide and 7,500 km (4,750 mi) long from Dakar to Djibouti.

Canada

In Canada, overall forest cover is increasing over the last few decades.

China

In China, extensive replanting programs have existed since the 1970s. Programs have had overall success. The forest cover has increased from 12% of China's land area to 16%. However, specific programs have had limited success. The "Green Wall of China", an attempt to limit the expansion of the Gobi Desert is planned to be 2,800 miles (4,500 km) long and to be completed in 2050. China plans to plant 26 billion trees in the next decade that is two trees for every Chinese citizen per year. China requires that students older than 11 years old plant one tree a year until their high school graduation.

Germany

Reforestation is required as part of the federal forest law. 31% of Germany is forested, according to the second forest inventory of 2001–2003. The size of the forest area in Germany increased between the first and the second forest inventory due to forestation of degenerated bogs and agricultural areas.

India

Jadav Payeng had received national awards for reforestation efforts, known as the "Molai forest".

Japan

Mass environmental and human-body pollution along with relating deforestation, water pollution, smoke damage, and loss of soils caused by mining operations in Ashio, Tochigi became the first environmental social issue in Japan, efforts by Shōzō Tanaka had grown to large campaigns against copper operation. This led to the creation of 'Watarase Yusuichi Pond', to settle the pollution which is a Ramsar site today. Reforestation was conducted as a part of afforestation due to inabilities of self-recovering by the natural land itself due to serious soil pollution and loss of woods consequence in loss of soils for plants to grow, thus needing artificial efforts involving introducing of healthy soils from outside. Starting since in 1897, about 50% of once bald mountains backed to green.

Lebanon

For thousands of years Lebanon was covered by forests, one particular species of interest, *Cedrus libani* or Lebanon Cedar was exceptionally valuable timber species. Virtually every ancient culture that shared the Mediterranean Sea harvested these trees, the Ancient Egyptians, the Greeks, the Roman Empire, early Christians, the Muslims, Persians, Assyrians and Babylonians. Extensive deforestation has occurred, with only small remnants of the original forests surviving. Deforestation has been particularly severe in Lebanon and on Cyprus. The Lebanon Reforestation Initiative aims to restore Lebanon's native forests. Projects financed locally and by international charity are performing extensive reforestation of cedar being carried out in the Mediterranean region, particularly in Lebanon and Turkey, where over 50 million young cedars are being planted annually.

Philippines

The Philippines increased its forest area to 240,000 hectares (593,053 acre) per year from 2010 to 2015. In 2011-2015 period, reforestation through the government's National Greening Program covered 1,258,692 hectares of land, which is 105 percent more than the 1,200,000 hectares planting target for the same period. An estimated eight million hectares of land nationwide need reforestation and rehabilitation, according to the Forest Management Bureau.

United States

It is the stated goal of the US Forest Service to manage forest resources sustainably. This includes reforestation after timber harvest, among other programs.

Organizations

Trees for the Future has assisted more than 170,000 families, in 6,800 villages of Asia, Africa and the Americas, to plant over 35 million trees.

Wangari Maathai, 2004 Nobel Peace Prize recipient, founded the Green Belt Movement which planted over 47 million trees to restore the Kenyan environment.

Shanghai Roots & Shoots, a division of the Jane Goodall Institute, launched The Million Tree Project in Kulun Qi, Inner Mongolia to plant one million trees to stop desertification and alleviate global warming.

Criticisms

Competition with Human Interests

A farmland near the Bennets Woodland, near Stichens Green and Streatley, Bedfordshire, the United Kingdom, 2005.

Reforestation competes with other land uses such as food production, livestock grazing and living space for further economic growth. However, in the case of food production, data from the Food and Agriculture Organization shows that between 1961 and 2012, the amount of arable land needed to produce the same amount of food declined by 68 percent worldwide.

Reforestation often has the tendency to create large fuel loads, resulting in significantly hotter combustion than fires involving low brush or grasses. It can divert large amounts of water from other activities. It sometimes results in extensive canopy creation that prevents growth of diverse vegetation in the shadowed areas and generating soil conditions that hamper other types of vegetation. Trees used in some reforesting efforts (e.g., *Eucalyptus globulus*) tend to extract large amounts of moisture from the soil, preventing the growth of other plants. An increase in the number of humans globally means that more food would have to be produced from farms to feed them. However, the growth rate is currently declining.

Natural Risks

There is also the risk that through a forest fire or insect outbreak much of the stored carbon in a reforested area could make its way back to the atmosphere. Reduced harvesting rates and fire suppression have caused an increase in the forest biomass in the western part of the Continental United States, over the past century. This causes an increase of about a factor of four in the frequency of fires due to longer and hotter dry seasons.

Urban Forestry

Tree pruning in Durham, North Carolina.

James Kinder, an ISA Certified Municipal Arborist examining a Japanese Hemlock at Hoyt Arboretum.

Urban forestry is the care and management of single trees and tree populations in urban settings for the purpose of improving the urban environment. Urban forestry advocates the role of trees as a critical part of the urban infrastructure. Urban foresters plant and maintain trees, support appropriate tree and forest preservation, conduct research and promote the many benefits trees provide. Urban forestry is practiced by municipal and commercial arborists, municipal and utility foresters, environmental policymakers, city planners, consultants, educators, researchers and community activists.

Functions and Values

Function, the dynamic operation of the forest, includes biochemical cycles, gas exchange, primary productivity, competition, succession, and regeneration. In urban environments, forest functions are frequently related to the human environment. Trees are usually selected, planted, trimmed, and nurtured by people, often with specific intentions, as when a tree is planted in a front yard to shade the driveway and frame the residence. The functional benefits provided by this tree depend on structural attributes, such as species and location, as well as management activities that influence its growth, crown dimensions, and health.

Urban forest functions are thus often oriented toward human outcomes, such as shade, beauty, and privacy. As prominent "things," arranged in distinctive formations, trees command a symbolic and material presence that informs how places and landscapes are imagined. This link that humans have to trees has been theorized by Kellert and Wilson (1993) to be a genetically based emotional need to be close to trees and other greenery. According to their "Biophilia Hypothesis," millions of years of human survival and evolution depended on our ability to cope with the natural world; learning what was safe and dangerous involved the imprinting of strong positive and negative emotional reactions to various natural stimuli. Although 21st century American society is no longer as dependent on nature for day-to-day survival, Kellert and Wilson suggest that closeness to the natural world is still critical for psychological well-being. The complex symbolic and emotional ties that humans have with trees have important implications for the importance of sound urban forest management practices that impact not only quality of life on an ecological level, but on the human and cultural level. People develop emotional attachments to trees that give them special status and value. Removing "hazardous" trees can be difficult when it means severing the connection between residents and the trees they love. For many, feelings of attachment to trees in cities influences feelings for preservation of trees in forests (McPherson 1998).

Professional Tree Climber (arborist: Zack Weiler) climbing a willow tree in Port Elgin, ON. Canada.

The value that people place on trees is especially evident with respect to big trees. There has always been a public fascination with large trees, especially the largest specimens of trees that reach a mature height of greater than 40 or 50 feet (i.e., Champion Trees) (Barro et al. 1997, Dwyer et al. 1991). Moreover, the ability of big street trees to create a ceiling of branches and leaves over all or part of a street impacts the scale of changing shadows cast by the trees, sunlight filtration, and other human-scale considerations that provide a changing visual environment (Zube 1973, Jones and Cloke 2002). In their qualitative study of Denmark residents' perceptions of the importance of the urban forest, Hansen-Moller and Oustrup (2004) found that the scale of urban trees was one of the main conditions of an "ideal" urban forest, through its volume, height, and ability to envelop a person, thus creating a barrier from the outside world.

Urban forests bring many environmental and economic benefits to cities. Among these are energy benefits in the form of reduced air conditioning by shading buildings, homes and roads, absorbing sunlight, reducing ultraviolet light, cooling the air, and reducing wind speed - in short improvement of the microclimate and air quality (McPherson 1994; McPherson & Rowntree 1993; Simpson & McPherson 1996; Coder 1996; Wolfe 1999; Hastie 2003; Lohr *et al.* 2004). There are also economic benefits associated with urban trees such as increased land, property, and rental value (Morales et al. 1983; Anderson & Cordell 1988; Wolf 1998; Dwyer et al. 1992; Mansfield et al. 2005; Orland et al. 1992; Hastie 2003; USDA Forest Service 2003, 2004). Well-maintained trees and landscaped business districts have been shown to encourage consumer purchases and attract increased residential, commercial and public investments (Wolf 2004, 2007). Trees located in business areas may also increase worker productivity, recruitment, retention and satisfaction (Kaplan & Kaplan 1989; Kaplan 1992; Wolf 1998). Urban forests also improve air quality, absorb rainwater, improve biodiversity and potentially allow recycling to 20% of waste which is wood-based

Many cities today are dealing with stormwater management system issues where their existing systems can no longer hold the volume of water that falls in storms. One sustainable solution to this is planting street trees with grates underneath them to hold water. Trees and their soils work to filter runoff pollution and soil contaminants by absorbing them and processing them into less harmful substances. They also collect water in their limbs and release it back into the atmosphere over time. This makes trees a solution to stormwater runoff issues and urban heating issues.

The social and even medical benefits of nature are also dramatic. Urban poverty is common to areas lacking green spaces. Visiting green areas in cities can counteract the stress of city life, renew vital energy and restore attention, and improve medical outcomes. Simply being able to see a natural view out of the window improves self-discipline in inner city girls.

Having regular access to woodland is desirable for schools, and indeed Forest kindergartens take children to visit substantial forests every day, whatever the weather. When such children go to primary school, teachers observe a significant improvement in reading, writing, mathematics, social skills and many other areas.

Various methods are available to capture the value of urban trees, each designed to analyse a specific type of green space (individual trees, parks, trees on golf courses etc.). The following are examples of studies that have used these different approaches, along with their respective constraints.

Method	Study	Location	Results	Limitations
Contingent valuation	Tyrvainen (2001)	Joensuu and Salo, Finland	More than two-thirds of the respondents were willing to pay for the use of recreation areas, with mean WTP ranging from 42 to 53 FIM/ month, depending on their location.	Estimated value of environmental amenities is based on a hypothetical market scenario
Choice modeling and survey	Salazar and Menendez (2007)	Valencia, Spain	Residents closer to a proposed park had a higher WTP for the park than those further from it.	Bias, protest answers, strategic answers
Direct estimates	Pandit and Laband (2010)	Auburn, Alabama, USA	17.5 percent tree cover on property = 14.4 percent reduction in electricity ($31/month) 50 percent dense shade = 19.3 percent reduction in electricity ($42/month).	Mitigation effects of climate excluded, Aesthetic values excluded
Numerical modeling	McPherson et al. (2005)	USA	Every dollar invested in urban tree management returned annual benefits ranging from $1.37 to $3.09.	Aesthetic values excluded

Practice

Urban forestry is a practical discipline, which includes tree planting, care, and protection, and the overall management of trees as a collective resource. The urban environment can present many arboricultural challenges such as limited root and canopy space, poor soil quality, deficiency or excess of water and light, heat, pollution, mechanical and chemical damage to trees, and mitigation of tree-related hazards. Among those hazards are mostly non-immediate risks like the probability that individual trees will not withstand strong winds (as during a thunderstorm) and damage parking cars or injure passing pedestrians. Although quite striking in an urban environment, large trees in particular present a continuing dilemma for the field of urban forestry due to the stresses that urban trees undergo from automobile exhaust, constraining hardscape and building foundations, and physical damage (Pickett et al. 2008). Urban forestry also challenges the arborists that tend the trees. The lack of space requires greater use of rigging skills and traffic and pedestrian control. The many constraints that the typical urban environment places on trees limits the average lifespan of a city tree to only 32 years – 13 years if planted in a downtown area – which is far short of the 150-year average life span of trees in rural settings (Herwitz 2001).

Management challenges for urban forestry include maintaining a tree and planting site inventory, quantifying and maximizing the benefits of trees, minimizing costs, obtaining and maintaining public support and funding, and establishing laws and policies for trees on public and on private land. Urban forestry presents many social issues that require addressing to allow urban forestry to be seen by the many as an advantage rather than a curse on their environment. Social issues include under funding which leads to inadequate maintenance of urban trees. In the UK the National Urban Forestry Unit produced a series of case studies around best practice in urban forestry which is archived here.

By Country

United States

Tree warden laws in the New England states are important examples of some of the earliest and most far-sighted state urban forestry and forest conservation legislation. In 1896, the Massachusetts legislature passed the first tree warden law, and the other five New England states soon followed suit: Connecticut, Rhode Island, and New Hampshire in 1901, Vermont in 1904, and Maine in 1919. (Kinney 1972, Favretti 1982, Campanella 2003).

As villages and towns grew in population and wealth, ornamentation of public, or common, spaces with shade trees also increased. However, the ornamentation of public areas did not evolve into a social movement until the late 18th century, when private individuals seriously promoted and sponsored public beautification with shade and ornamental trees (Favretti 1982, Lawrence 1995). Almost a century later, around 1850, institutions and organization were founded to promote ornamentation through private means (Egleston 1878, Favretti 1982). In the 1890s, New England's "Nail" laws enabled towns to take definitive steps to distinguish which shade trees were public. Chapter 196 of the 1890 Massachusetts Acts and Resolves stated that a public shade tree was to be designated by driving a nail or spike, with the letter M plainly impressed on its head, into the relevant trunk. Connecticut passed a similar law in 1893, except its certified nails and spikes bore the letter C. (Northrup 1887).

The rapid urbanization of American cities in the late 19th century was a concern to many as encouraging intellectual separation of humanity and nature (Rees 1997). By the end of the 19th century, social reformers were just beginning to understand the relationship between developing parks in urban areas and "[engendering] a better society" (Young 1995:536). At this time, parks and trees were not necessarily seen as a way to allow urban dwellers to experience nature, but more of a means of providing mechanisms of acculturation and control for newly arrived immigrants and their children (e.g., areas to encourage "structured play" and thus serve as a deterrent for youth crime) (Pincetl and Gearin 2005). Other prominent public intellectuals were interested in exploring the synergy between ecological and social systems, including American landscape architect Fredrick Law Olmsted, designer of 17 major U.S. urban parks and a visionary in seeing the value of including green space and trees as a fundamental part of metropolitan infrastructure (Young 2009). To Olmsted, unity between nature and urban dwellers was not only physical, but also spiritual: "Gradually and silently the charm comes over us; the beauty has entered our souls; we know not exactly when or how, but going away we remember it with a tender, subdued, filial-like joy" (Beveridge and Schuyler 1983 cited in Young 2009:320). The conscious inclusion of trees in urban designs for American cities such as Chicago, San Francisco, and Minneapolis was also inspired by Paris's urban forest and its broad, tree-lined boulevards as well as by the English romantic landscape movement (Zube 1973). The belief in green cover by early park proponents as a promoter of social cohesion has been corroborated by more recent research that links trees to the presence of stronger ties among neighbors, more adult supervision of children in outdoor areas, more use of the neighborhood common areas, and fewer property and violent crime (Kuo et al. 1998, Kuo and Sullivan 2001, Kuo 2003).

Many municipalities throughout the United States employ community-level tree ordinances to empower planning officials to regulate the planting, maintenance, and preservation of trees. The development of tree ordinances emerged largely as a response to the Dutch Elm Disease that plagued cities from the 1930s to 1960s, and grew in response to urban development, loss of urban tree canopy, and rising public concern for the environment (Wolf 2003). The 1980s saw the beginning of the second generation of ordinances with higher standards and specific foci, as communities sought to create more environmentally pleasing harmony between new development and existing infrastructure. These new ordinances, legislated by local governments, may include specific provisions such as the diameter of tree and percentage of trees to be protected during construction activities (Xiao 1995). The implementation of these tree ordinances is greatly aided by a significant effort by community tree advocates to conduct public outreach and education aimed at increasing environmental concern for urban trees, such as through National Arbor Day celebrations and the USDA Urban and Community Forestry Program (Dwyer et al. 2000, Hunter and Rinner 2004, Norton and Hannon 1997, Wall et al. 2006). Much of the work on the ground is performed by non-profits funded by private donations and government grants.

Policy on urban forestry is less contentious and partisan than many other forestry issues, such as resource extraction in national forests. However, the uneven distribution of healthy urban forests across the landscape has become a growing concern in the past 20 years. This is because the urban forest has become an increasingly important component of bioregional ecological health with the expanding ecological footprint of urban areas. Based on American Forests' Urban Ecosystem Analyses conducted over the past six years in ten cities, an estimated 634,407,719 trees have been lost from metropolitan areas across the U.S. as the result of urban and suburban development (American Forests 2011). This is often due to the failure of municipalities to integrate trees and other elements of the green infrastructure into their day-to-day planning and decision-making processes (American Forests 2002). The inconsistent quality of urban forestry programs on the local level ultimately impacts the regional context in which contiguous urban forests reside, and is greatly exacerbated by suburban sprawl as well as other social and ecological effects (Webb et al. 2008). The recognition of this hierarchical linkage among healthy urban forests and the effectiveness of broader ecosystem protection goals (e.g., maintaining biodiversity and wildlife corridors), highlights the need for scientists and policymakers to gain a better understanding of the socio-spatial dynamics that are associated with tree canopy health at different scales (Wu 2008).

United Kingdom

In the UK urban forestry was pioneered around the turn of the 19th century by the Midland reafforesting association, whose focus was in the Black Country. In the mid 1990s the National Urban Forestry Unit (NUFU) grew out of a Black Country Urban Forestry Unit and promoted urban forestry across the UK, notably including the establishment of the Black Country Urban Forest. As urban forestry become more mainstream in the 21st century, NUFU was wound up, and its advocacy role now carried on by organisations such as The Wildlife Trusts and the Woodland Trust.

Toronto

Toronto is a diverse city with a mosaic urban forest – a patchwork of unique situations made up of trees growing in the many residential yards, lining the public streets, and beautifying public parks. Unlike the trees that grow in a wild setting, urban trees are faced with harsh conditions that can be detrimental to their health and growing potential. Soil compaction, air pollution, habitat fragmentation and competition from invasive species are some of the hardships city trees endure. Some neighbourhoods have a geriatric tree population; many mature trees that will reach the end of their lifespan very soon, with few young trees to replace them.

Some neighbourhoods suffer a serious lack of species diversity, with mainly ornamental, non-native or invasive tree species such as Bradford pear, Japanese tree lilac and Norway maple. Still other neighbourhoods, most often newly constructed subdivisions, lack tree cover completely.

Simply planting more trees cannot solve the problems faced by the urban forest. Through creative and innovative approaches the public, partnered with private enterprises can maximize the potential benefits of trees planted, and minimize the stresses they will have to overcome.

Although most people express a concern for urban trees and consider them very important, many lack the basic knowledge and skills needed to address and prevent the issues listed above. Collective action, or inaction, will make or break the future of the urban forest. Through fostering a sense of ownership amongst Toronto residents for this commonly owned resource, residents will enjoy better air quality and reduce their demand for energy.

Constraints

Resolving limitations will require coordinated efforts among cities, regions, and countries (Meza, 1992; Nilsson, 2000; Valencia, 2000).

- Loss of green space is continuous as cities expand; available growing space is limited in city centres. This problem is compounded by pressure to convert green space, parks, etc. into building sites (Glickman, 1999).
- Inadequate space is allowed for the root system.
- Poor soil is used when planting specimens.

- Incorrect and neglected staking leads to bark damage.
- Larger, more mature trees are often used to provide scale and a sense of establishment to a scheme. These trees grow more slowly and do not thrive in alien soils whilst smaller specimens can adapt more readily to existing conditions.
- Lack of information on the tolerances of urban tree cultivars to environmental constraints.
- Poor tree selection which leads to problems in the future.
- Poor nursery stock and failure of post-care.
- Limited genetic diversity.
- Too few communities have working tree inventories and very few have urban forest management plans.
- Lack of public awareness about the benefits of healthy urban forests.
- Poor tree care practices by citizens and untrained arborists.

Organizations

- Alliance for Community Trees.
- American Forests.
- Casey Trees.
- Friends of the Urban Forest.
- Greening of Detroit.
- Hantz Woodlands.
- International Society of Arboriculture.
- National Urban Forestry Unit.
- OpenTreeMap.
- Our City Forest.
- Society of Municipal Arborists.
- Society of American Foresters.
- LEAFTennessee Urban Forestry Council.
- The Tree Council, UK.
- Tree City USA Program.
- TREE Fund.
- TreeLink.
- Trees Are Good.
- Canopy.org.
- Trees Atlanta.
- Arborist Video Blog.
- TreesCharlotte.

Biosequestration

Biosequestration is the capture and storage of the atmospheric greenhouse gas carbon dioxide by biological processes.

This may be by increased photosynthesis (through practices such as reforestation / preventing deforestation and genetic engineering); by enhanced soil carbon trapping in agriculture; or by the use of algal bio sequestration to absorb the carbon dioxide emissions from coal, petroleum (oil) or natural gas-fired electricity generation.

Flowering Corymbia ficifolia, Austins Ferry, Tasmania, Australia.

Biosequestration as a natural process has occurred in the past, and was responsible for the formation of the extensive coal and oil deposits which are now being burned. It is a key policy concept in the climate change mitigation debate. It does not generally refer to the sequestering of carbon dioxide in oceans or rock formations, depleted oil or gas reservoirs, deep saline aquifers, or deep coal seams or through the use of industrial chemical carbon dioxide scrubbing.

The importance of Plants in Storing Atmospheric Carbon Dioxide

Kew Gardens Waterlily House. David Iliff, 2008.

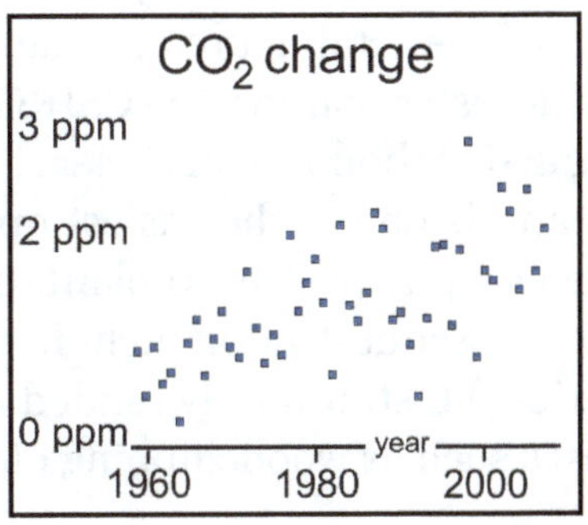

Recent year-to-year increase of atmospheric CO_2.

After water vapour (concentrations of which humans have limited capacity to influence) carbon dioxide is the most abundant and stable greenhouse gas in the atmosphere (methane rapidly reacts to form water vapour and carbon dioxide). Atmospheric carbon dioxide has increased from about 280 ppm in 1750 to 383 ppm in 2007 and is increasing at an average rate of 2 ppm pr year. The world's oceans have previously played an important role in sequestering atmospheric carbon dioxide through solubility and the action of phytoplankton. This, and the likely adverse consequences for humans and the biosphere of associated global warming, increases the significance of investigating policy mechanisms for encouraging biosequestration.

Reforestation, Avoided Deforestation and LULUCF

Reforestation and reducing deforestation can increase biosequestration in four ways. Pandani (Richea pandanifolia) near Lake Dobson, Mount Field National Park, Tasmania, Australia.

The Intergovernmental Panel on Climate Change (IPCC) estimates that the cutting down of forests is now contributing close to 20 per cent of the overall greenhouse gases entering the atmosphere. Candell and Raupach argue that there are four primary ways in which reforestation and reducing deforestation can increase biosequestration. First, by increasing the volume of existing forest. Second, by increasing the carbon density of existing forests at a stand and landscape scale. Third, by expanding the use of forest products that will sustainably replace fossil-fuel emissions. Fourth, by reducing carbon emissions that are caused from deforestation and degradation. Land clearing reductions, the majority of the time, create biodiversity benefits in a vast expanse of land regions. Concerns, however, arise when the density and area of vegetation increases the grazing pressure could also increase in other areas, causing land degradation.

A recent report by the Australian CSIRO found that forestry and forest-related options are the most significant and most easily achieved carbon sink making up 105 Mt per year CO_2-e or about 75 per cent of the total figure attainable for the Australian state of Queensland from 2010-2050. Among the forestry options, the CSIRO report announced, forestry with the primary aim of carbon storage (called carbon forestry) clearly has the highest attainable carbon storage capacity (77 Mt CO_2-e/yr) and is one of the easiest options to implement compared with biodiversity plantings, pre-1990 eucalypts, post 1990 plantations and managed regrowth. Legal strategies to encourage this form of biosequestration include permanent protection of forests in National Parks or on the World Heritage List, properly funded management and bans on use of rainforest timbers and inefficient uses such as woodchipping old growth forest.

As a result of lobbying by the developing country caucus (or Group of 77) in the United Nations (associated with the United Nations Conference on Environment and Development (UNCED) in Rio de Janeiro, the non-legally binding Forest Principles were established in 1992. These linked the problem of deforestation to third world debt and inadequate technology transfer and stated that the "agreed full incremental cost of achieving benefits associated with forest conservation-should be equitably shared by the international community" (para1(b)). Subsequently, the Group of 77 argued in the 1995 *Intergovernmental Panel on Forests* (IPF) and then the 2001 *Intergovernmental Forum on Forests* (IFF), for affordable access to environmentally sound technologies without the stringency of intellectual property rights; while developed states there rejected demands for a forests fund. The expert group created under the United Nations Forum on Forests

(UNFF) reported in 2004, but in 2007 developed nations again vetoed language in the principles of the final text which might confirm their legal responsibility under international law to supply finance and environmentally sound technologies to the developing world.

Settlement and deforestation surrounding the Brazilian town of Rio Branco are seen here in the striking "herring bone" deforestation patterns that cut through the rainforest. NASA, 2008.

In December 2007, after a two-year debate on a proposal from Papua New Guinea and Costa Rica, state parties to the United Nations Framework Convention on Climate Change (FCCC) agreed to explore ways of reducing emissions from deforestation and to enhance forest carbon stocks in developing nations. The underlying idea is that developing nations should be financially compensated if they succeed in reducing their levels of deforestation (through valuing the carbon that is stored in forests); a concept termed 'avoided deforestation (AD) or, REDD if broadened to include reducing forest degradation. Under the free market model advocated by the countries who have formed the *Coalition of Rainforest Nations*, developing nations with rainforests would sell carbon sink credits under a free market system to Kyoto Protocol Annex I states who have exceeded their emissions allowance. Brazil (the state with the largest area of tropical rainforest) however, opposes including avoided deforestation in a carbon trading mechanism and instead favors creation of a multilateral development assistance fund created from donations by developed states. For REDD to be successful science and regulatory infrastructure related to forests will need to increase so nations may inventory all their forest carbon, show that they can control land use at the local level and prove that their emissions are declining.

NASA Earth Observatory, 2009. Deforestation in Malaysian Borneo.

Subsequent to the initial donor nation response, the UN established REDD Plus, or REDD+, expanding the original program's scope to include increasing forest cover through both reforestation and the planting of new forest cover, as well as promoting sustainable forest resource management.

The United Nations Framework Convention on Climate Change (UNFCCC) Article 4(1)(a) requires all Parties to "develop, periodically update, publish and make available to the Conference of the Parties" as well as "national inventories of anthropogenic emissions by sources" "removals by sinks of all greenhouse gases not controlled by the Montreal Protocol." Under the UNFCCC reporting guidelines, human-induced greenhouse emissions must be reported in six sectors: energy (including stationary energy and transport); industrial processes; solvent and other product use; agriculture; waste; and *land use, land use change and forestry* (LULUCF). The rules governing accounting and reporting of greenhouse gas emissions from LULUCF under the Kyoto Protocol are contained in several decisions of the Conference of Parties under the UNFCCC and LULUCF has been the subject of two major reports by the Intergovernmental Panel on Climate Change (IPCC). The Kyoto Protocol article 3.3 thus requires mandatory LULUCF accounting for afforestation (no forest for last 50 years), reforestation (no forest on 31 December 1989) and deforestation, as well as (in the first commitment period) under article 3.4 voluntary accounting for cropland management, grazing land management, revegetation and forest management (if not already accounted under article 3.3).

Continent of Australia from space. Australia is a major producer of fossil fuels and has significant problems with deforestation.

As an example, the *Australian National Greenhouse Gas Inventory* (NGGI) prepared in compliance with these requirements indicates that the energy sector accounts for 69 per cent of Australia's emissions, agriculture 16 per cent and LULUCF six per cent. Since 1990, however, emissions from the energy sector have increased 35 per cent (stationary energy up 43% and transport up 23%). By comparison, emissions from LULUCF have fallen by 73%. However, questions have been raised by Andrew Macintosh about the veracity of the estimates of emissions from the LULUCF sector because of discrepancies between the Australian Federal and Queensland Governments' land clearing data. Data published by the *Statewide Landcover and Trees Study* (SLATS) in Queensland, for example, show that the total amount of land clearing in Queensland identified under SLATS between 1989/90 and 2000/01 is approximately 50 per cent higher than the amount estimated by the Australian Federal Government's *National Carbon Accounting System* (NCAS) between 1990 and 2001.

Deforestation in Haiti. NASA, 2008.

Satellite imaging has become crucial in obtaining data on levels of deforestation and reforestation. Landsat satellite data, for example, has been used to map tropical deforestation as part of NASA's Landsat *Pathfinder Humid Tropical Deforestation Project*, a collaborative effort among scientists from the University of Maryland, the University of New Hampshire, and NASA's Goddard Space Flight Center. The project yielded deforestation maps for the Amazon Basin, Central Africa, and Southeast Asia for three periods in the 1970s, 1980s, and 1990s.

Enhanced Photosynthesis

Sprekelia formosissima in Tasmania, Australia.

Biosequestration may be enhanced by improving photosynthetic efficiency by modifying RuBisCO genes in plants to increase the catalytic and/or oxygenation activity of that enzyme.

One such research area involves increasing the Earth's proportion of C4 carbon fixation photosynthetic plants. C4 plants represent about 5% of Earth's plant biomass and 1% of its known plant species, but account for around 30% of terrestrial carbon fixation. In leaves of C3 plants, captured photons of solar energy undergo photosynthesis which assimilates carbon into carbohydrates (triosephosphates) in the chloroplasts of the mesophyll cells. The primary CO_2 fixation step is catalysed by ribulose-1,5-bisphosphate carboxylase/oxygenase (Rubisco) which reacts with O2 leading to photorespiration that protects photosynthesis from photoinhibition but wastes 50% of potentially fixed carbon. The C4 photosynthetic pathway, however, concentrates CO_2 at the site of the reaction of Rubisco, thereby reducing the biosequestration-inhibiting photorespiration. A new frontier in crop science consists of attempts to genetically engineer C3 staple food crops (such as wheat, barley, soybeans, potatoes and rice) with the "turbo-charged" photosynthetic apparatus of C4 plants.

Hakea epiglottis, Cape Raoul, Tasman Peninsula, Tasmania, Australia.

Biochar

Biochar (charcoal created by pyrolysis of biomass) is a potent form of longterm (thousands of years) biosequestration of atmosphereic CO_2 derived from investigation of the extremely fertile Terra preta soils of the Amazon Basin. Placing biochar in soils also improves water quality, increases soil fertility, raises agricultural productivity and reduce pressure on old growth forests. As a method of generating bio-energy with carbon storage Rob Flanagan and the EPRIDA biochar company have developed low-tech cooking stoves for developing nations that can burn agricultural wastes such as rice husks and produce 15% by weight of biochar; while BEST Energies in NSW Australia have spent a decade developing an Agrichar technology that can combust 96 tonnes of dry biomass each day, generating 30-40 tonnes of biochar. A parametric study of biosequestration by Malcolm Fowles at the Open University, indicated that to mitigate global warming, policies should encourage displacement of coal with biomass as a power source for baseload electricity generation if the latter's conversion efficiency rose over 30%, otherwise *biosequestering* carbon from biomass as a cheaper mitigation option than geosequestration by CO_2 capture and storage.

Improved Agricultural and Farming Practices

Zero-till farming practices occur where there is much mulching but ploughing is not used, so that the carbon-rich organic matter in soil is not exposed to atmospheric oxygen, or to the leaching and erosion effects of rainfall. Ceasing ploughing has been alleged to encourage more ants to become predators of wood-eating (and CO_2 generating) termites, allows weeds to regenerate soils and helps slow water flows over the land.

Shepherds with their sheep.

Soil holds more carbon than vegetation and atmosphere combined, and most soil lies under grazing land. Holistic Planned Grazing holds tremendous potential in mitigating global warming, while building soil, increasing biodiversity, and reversing desertification. Developed by Allan Savory, it uses fencing and/or herders, to restore grasslands by carefully planning movements of large herds of livestock to mimic the vast herds found in nature where grazing animals are kept concentrated by pack predators and forced to move on after eating, trampling, and manuring an area, returning only after it has fully recovered. This method of grazing seeks to emulate what occurred during the past 40 million years as the expansion of grass-grazer ecosystems built deep, rich grassland soils, sequestering carbon and cooling the planet.

Dedicated biofuel and biosequestration crops, such as switchgrass (panicum virgatum), are also being developed. It requires from 0.97 to 1.34 GJ fossil energy to produce 1 tonne of switchgrass, compared with 1.99 to 2.66 GJ to produce 1 tonne of corn. Given that switchgrass contains approximately 18.8 GJ/ODT of biomass, the energy output-to-input ratio for the crop can be up to 20:1.

Panicum virgatum switchgrass, valuable in biofuel production, soil conservation and biosequestration.

Biosequestration can also be enhanced by farmers choosing crops species that produce large numbers of phytoliths. Phytoliths are microscopic spherical shells of silicon that can store carbon for thousands of years.

Biosequestration and Climate Change Policy

Biosequestration could be critical to climate change mitigation till cleaner forms of power generation are established. The Nesjavellir Geothermal Power Plant in Þingvellir, Iceland.

Industries with large amounts of CO_2 emissions (such as the coal industry) are interested in biosequestration as a means of offsetting their greenhouse gas production. In Australia, university researchers are engineering algae to produce biofuels (hydrogen and biodiesel oils) and investigating whether this process can be used to *biosequester* carbon. Algae naturally capture sunlight and use its energy to split water into hydrogen, oxygen and oil which can be extracted. Such clean energy production also can be coupled with desalination using salt-tolerant marine algae to generate fresh water and electricity.

Windturbines D4 (nearest) to D1 on the Thornton Bank.

Many new bioenergy (biofuel) technologies, including cellulosic ethanol biorefineries (using stems and branches of most plants including crop residues such as corn stalks, wheat straw and rice straw) are being promoted because they have the added advantage of biosequestration of CO_2. The Garnaut Climate Change Review recommends that a carbon price in a carbon emission trading scheme could include a financial incentive for biosequestration processes. Garnaut recommends the use of algal biosequestration to absorb the constant stream of carbon dioxide emissions from coal-fired electricity generation and metal smelting until renewable forms of energy, such as solar and wind power, become more established contributors to the grid. Garnaut, for example, states: "Some algal biosequestration processes could absorb emissions from coal-fired electricity generation and metals smelting." The United Nations Collaborative Programme on Reducing Emissions from Deforestation and Forest Degradation in Developing Countries (UN-REDD Programme) is a collaboration between FAO, UNDP and UNEP under which a trust fund established in July 2008 allows donors to pool resources to generate the requisite transfer flow of resources to significantly reduce global emissions from deforestation and forest degradation. The UK government's Stern Review on the economics of climate change argued that curbing deforestation was a "highly cost-effective way of reducing greenhouse gas emissions".

James E. Hansen argues that, "An effective way to achieve drawdown [of carbon dioxide] would be to burn biofuels in power plants and capture the CO_2, with the biofuels derived from agricultural or urban wastes or grown on degraded lands using little or no fossil fuel inputs." Such CO_2 drawdown systems are referred to as Bio-energy with carbon capture and storage, or BECCS. According to a study by Biorecro and the Global CCS Institute, there is currently (as of January 2012) 550 000 tonnes CO_2/year in total BECCS capacity operating, divided between three different facilities.

Under a 2009 agreement, Loy Yang Power and MBD Energy Ltd will build a pilot Fossil fuel power plant at the Latrobe Valley power station in Australia using biosequestration technology in the form of an algal synthesiser system. Captured CO_2 from the waste exhaust flue gases will be injected into circulating waste water to grow oil-rich algae where sunlight and nutrients will produce heavy oil-laden slurry that can make high grade oil for energy, or stock feed. Other commercial demonstration projects involving biosequestration of CO_2 at point of emission have begun in Australia.

Philosophical Basis of Biosequestration

The arguments for biosequestration are often shaped in terms of economic theory, yet there is a well-recognised quality of life dimension to this debate. Biosequestration assists human beings to increase their collective and individual contributions to the essential resources of the biosphere. The policy case for biosequestration overlaps with principles of ecology, sustainability and sustainable development, as well as biosphere, biodiversity and ecosystem protection, environmental ethics, climate ethics and natural conservation.

Barriers to Increased Global Biosequestration

Lassen National Park, Kings Creek, USA.

The Garnaut Climate Change Review notes many barriers to increased global biosequestration. "There must be changes in the accounting regimes for greenhouse gases. Investments are required in research, development and commercialisation of superior approaches to biosequestration. Adjustments are required in the regulation of land use. New institutions will need to be developed to coordinate the interests in utilisation of biosequestration opportunities across small business in rural communities. Special efforts will be required to unlock potential in rural communities in developing countries." Saddler and King have argued that biosequestration and agricultural greenhouse gas emissions should not be handled within a global emissions trading scheme because of difficulties with measuring such emissions, problems in controlling them and the burden that would be placed on numerous small-scale farming operations. Collett likewise maintains that REDD credits (post-facto payments to developing countries for reducing their deforestation rates below an historical or projected reference rate), simply create a complex market approach to this global public health problem that reduces transparency and accountability when targets are not met and will not be as effective as developed nations voluntarily funding countries to keep their rainforests.

The World Rainforest Movement has argued that poor developing countries could be pressured to accept reforestation projects under the Kyoto Protocol's Clean Development Mechanism in order to earn foreign exchange simply to pay off the interest on debt to the World Bank. Tensions also exist over forest management between the sovereignty claims of nations states, arguments about common heritage of mankind and the rights of indigenous peoples and local communities; the Forest Peoples Programme (FPP) arguing the anti-deforestation programs could merely allow financial benefits to flow to national treasuries, privilege would-be corporate forest degraders who manipulate the system by periodically threatening forests, rather than local communities who conserve them. The success of such projects will also depend on the accuracy of the baseline data and the number of countries involved. Further, it has been argued that if biosequestration is to play a significant role in mitigating anthropogenic climate change then coordinated policies should set a goal of achieving global forest cover to its extent prior to the industrial revolution in the 1800s.

It has also been argued that the United Nations mechanism for Reducing Emissions from Deforestation and Forest Degradation (REDD) may increase pressure to convert or modify other ecosystems, especially savannahs and wetlands, for food or biofuel, even though those ecosystems also have high carbon sequestration potential. Globally, for example, peatlands cover only 3% of the land surface but store twice the amount of carbon as all the world's forests, whilst mangrove forests and saltmarshes are examples of relatively low-biomass ecosystems with high levels of productivity and carbon sequestration. Other researchers have argued that REDD is a critical component of an effective global biosequestration strategy that could provide significant benefits, such as the conservation of biodiversity, particularly if it moves away from focusing on protecting forests that are most cost-effective for reducing carbon emissions (such as those in Brazil where agricultural opportunity costs are relatively low, unlike Asia, which has sizeable revenues from oil palm, rubber, rice, and maize). They argue REDD could be varied to allow funding of programs to slow peat degradation in Indonesia and target protection of biodiversity in "hot spot"—areas with high species richness and relatively little remaining forest. Some purchasers, they maintain, of REDD carbon credits, such as multinational corporations or nations, might pay a premium to save imperiled eco-systems or areas with high-profile species.

Carbon Sequestration

Carbon sequestration is the process involved in carbon capture and the long-term storage of atmospheric carbon dioxide. Carbon sequestration involves long-term storage of carbon dioxide or other forms of carbon to mitigate or defer global warming. It has been proposed as a way to slow the atmospheric and marine accumulation of greenhouse gases, which are released by burning fossil fuels.

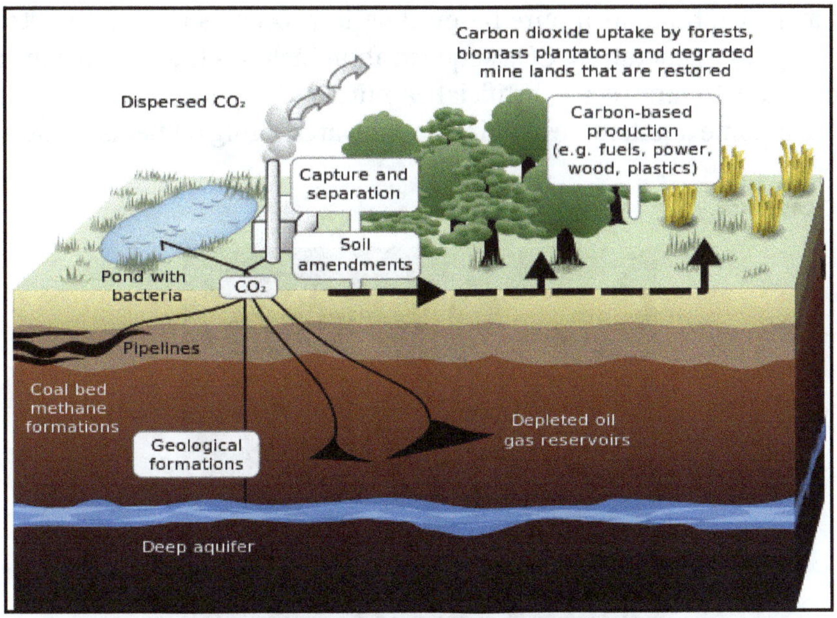

Schematic showing both terrestrial and geological sequestration of carbon dioxide emissions from a coal-fired plant.

Carbon dioxide (CO_2) is naturally captured from the atmosphere through biological, chemical, and physical processes. Artificial processes have been devised to produce similar effects, including large-scale, artificial capture and sequestration of industrially produced CO2 using subsurface saline aquifers, reservoirs, ocean water, aging oil fields, or other carbon sinks.

Description

Carbon sequestration is the process involved in carbon capture and the long-term storage of atmospheric carbon dioxide (CO_2) and may refer specifically to:

- "The process of removing carbon from the atmosphere and depositing it in a reservoir." When carried out deliberately, this may also be referred to as carbon dioxide removal, which is a form of geoengineering.
- Carbon capture and storage, where carbon dioxide is removed from flue gases (e.g., at power stations) before being stored in underground reservoirs.
- Natural biogeochemical cycling of carbon between the atmosphere and reservoirs, such as by chemical weathering of rocks.

Carbon sequestration describes long-term storage of carbon dioxide or other forms of carbon to either mitigate or defer global warming and avoid dangerous climate change. It has been proposed as a way to slow the atmospheric and marine accumulation of greenhouse gases, which are released by burning fossil fuels.

Carbon dioxide is naturally captured from the atmosphere through biological, chemical or physical processes. Some artificial sequestration techniques exploit these natural processes, while some use entirely artificial processes.

Carbon dioxide may be captured as a pure by-product in processes related to petroleum refining or from flue gases from power generation. CO_2 sequestration includes the storage part of carbon capture and storage, which refers to large-scale, artificial capture and sequestration of industrially produced CO_2 using subsurface saline aquifers, reservoirs, ocean water, aging oil fields, or other carbon sinks.

Biological Processes

An oceanic phytoplankton bloom in the South Atlantic Ocean, off the coast of Argentina. Encouraging such blooms with iron fertilization could lock up carbon on the seabed.

Biosequestration or carbon sequestration through biological processes affects the global carbon cycle. Examples include major climatic fluctuations, such as the Azolla event, which created the current Arctic climate. Such processes created fossil fuels, as well as clathrate and limestone. By manipulating such processes, geoengineers seek to enhance sequestration.

Peat Production

Peat bogs are a very important carbon store. Peat bogs act as a sink for carbon due to the accumulation of partially decayed biomass that would otherwise continue to decay completely. There is a variance on how much the peatlands act as a carbon sink or carbon source that can be linked to varying climates in different areas of the world and different times of the year. By creating new bogs, or enhancing existing ones, the amount of carbon that is sequestered by bogs would increase.

Forestry

Reforestation is the replanting of trees on marginal crop and pasture lands to incorporate carbon from atmospheric CO_2 into biomass. For this process to succeed the carbon must not return to the atmosphere from mass burning or rotting when the trees die. To this end, land allotted to the trees must not be converted to other uses and management of the frequency of disturbances might be necessary in order to avoid extreme events. Alternatively, the wood from them must itself be sequestered, e.g., via biochar, bio-energy with carbon storage (BECS), landfill or 'stored' by use in e.g. construction. Short of growth in perpetuity, however, reforestation with long-lived trees (>100 years) will sequester carbon for a more graduated release, minimizing impact during the expected carbon crisis of the 21st century.

Urban Forestry

Urban Forestry increases the amount of carbon taken up in cities by adding new tree sites and the sequestration of carbon occurs over the lifetime of the tree. It is generally practiced and maintained on smaller scales, like in cities. The results of urban forestry can have different results depending on the type of vegetation that is being used, so it can function as a sink but can also function as a source of emissions. Along with sequestration by the plants which is difficult to measure but seems to have little effect on the overall amount of carbon dioxide that is uptaken, the vegetation can have indirect effects on carbon by reducing need for energy consumption.

Wetland Restoration

Wetland soil is an important carbon sink; 14.5% of the world's soil carbon is found in wetlands, while only 6% of the world's land is composed of wetlands.

Agriculture

Globally, soils are estimated to contain approximately 1,500 gigatons of organic carbon to 1 m depth, more than the amount in vegetation and the atmosphere.

Modification of agricultural practices is a recognized method of carbon sequestration as soil can act as an effective carbon sink offsetting as much as 20% of 2010 carbon dioxide emissions annually.

Carbon emission reduction methods in agriculture can be grouped into two categories: reducing and/or displacing emissions and enhancing carbon removal. Some of these reductions involve increasing the efficiency of farm operations (e.g. more fuel-efficient equipment) while some involve interruptions in the natural carbon cycle. Also, some effective techniques (such as the elimination of stubble burning) can negatively impact other environmental concerns (increased herbicide use to control weeds not destroyed by burning).

Reducing Emissions

Increasing yields and efficiency generally reduces emissions as well, since more food results from the same or less effort. Techniques include more accurate use of fertilizers, less soil disturbance, better irrigation, and crop strains bred for locally beneficial traits and increased yields.

Replacing more energy intensive farming operations can also reduce emissions. Reduced or no-till farming requires less machine use and burns correspondingly less fuel per acre. However, no-till usually increases use of weed-control chemicals and the residue now left on the soil surface is more likely to release its CO_2 to the atmosphere as it decays, reducing the net carbon reduction.

In practice, most farming operations that incorporate post-harvest crop residues, wastes and by-products back into the soil provide a carbon storage benefit. This is particularly the case for practices such as field burning of stubble - rather than releasing almost all of the stored CO2 to the atmosphere, tillage incorporates the biomass back into the soil.

Enhancing Carbon Removal

All crops absorb CO_2 during growth and release it after harvest. The goal of agricultural carbon removal is to use the crop and its relation to the carbon cycle to permanently sequester carbon within the soil. This is done by selecting farming methods that return biomass to the soil and enhance the conditions in which the carbon within the plants will be reduced to its elemental nature and stored in a stable state. Methods for accomplishing this include:

- Use cover crops such as grasses and weeds as temporary cover between planting seasons.
- Concentrate livestock in small paddocks for days at a time so they graze lightly but evenly. This encourages roots to grow deeper into the soil. Stock also till the soil with their hooves, grinding old grass and manures into the soil.
- Cover bare paddocks with hay or dead vegetation. This protects soil from the sun and allows the soil to hold more water and be more attractive to carbon-capturing microbes.
- Restore degraded land, which slows carbon release while returning the land to agriculture or other use.

Agricultural sequestration practices may have positive effects on soil, air, and water quality, be beneficial to wildlife, and expand food production. On degraded croplands, an increase of 1 ton of soil carbon pool may increase crop yield by 20 to 40 kilograms per hectare of wheat, 10 to 20 kg/ha for maize, and 0.5 to 1 kg/ha for cowpeas.

The effects of soil sequestration can be reversed. If the soil is disrupted or tillage practices are abandoned, the soil becomes a net source of greenhouse gases. Typically after 15 to 30 years of sequestration, soil becomes saturated and ceases to absorb carbon. This implies that there is a global limit to the amount of carbon that soil can hold.

Many factors affect the costs of carbon sequestration including soil quality, transaction costs and various externalities such as leakage and unforeseen environmental damage. Because reduction of atmospheric CO_2 is a long-term concern, farmers can be reluctant to adopt more expensive agricultural techniques when there is not a clear crop, soil, or economic benefit. Governments such as Australia and New Zealand are considering allowing farmers to sell carbon credits once they document that they have sufficiently increased soil carbon content.

Ocean-related

Iron Fertilization

Ocean iron fertilization is an example of such a geoengineering technique. Iron fertilization attempts to encourage phytoplankton growth, which removes carbon from the atmosphere for at least a period of time. This technique is controversial due to limited understanding of its complete effects on the marine ecosystem, including side effects and possibly large deviations from expected behavior. Such effects potentially include release of nitrogen oxides, and disruption of the ocean's nutrient balance.

Natural iron fertilisation events (e.g., deposition of iron-rich dust into ocean waters) can enhance carbon sequestration. Sperm whales act as agents of iron fertilisation when they transport iron from the deep ocean to the surface during prey consumption and defecation. Sperm whales have been shown to increase the levels of primary production and carbon export to the deep ocean by depositing iron rich feces into surface waters of the Southern Ocean. The iron rich feces causes phytoplankton to grow and take up more carbon from the atmosphere. When the phytoplankton dies, some of it sinks to the deep ocean and takes the atmospheric carbon with it. By reducing the abundance of sperm whales in the Southern Ocean, whaling has resulted in an extra 200,000 tonnes of carbon remaining in the atmosphere each year.

Urea Fertilization

Ian Jones proposes fertilizing the ocean with urea, a nitrogen rich substance, to encourage phytoplankton growth.

Australian company Ocean Nourishment Corporation (ONC) plans to sink hundreds of tonnes of urea into the ocean to boost CO_2-absorbing phytoplankton growth as a way to combat climate change. In 2007, Sydney-based ONC completed an experiment involving 1 tonne of nitrogen in the Sulu Sea off the Philippines.

Mixing Layers

Encouraging various ocean layers to mix can move nutrients and dissolved gases around, offering avenues for geoengineering. Mixing may be achieved by placing large vertical pipes in the oceans to pump nutrient rich water to the surface, triggering blooms of algae, which store carbon when they grow and export carbon when they die. This produces results somewhat similar to iron fertilization. One side-effect is a short-term rise in CO_2, which limits its attractiveness.

Seaweed

Seaweed grows very fast and can theoretically be harvested and processed to generate biomethane, via Anaerobic Digestion to generate electricity, via Cogeneration/CHP or as a replacement for natural gas. One study suggested that if seaweed farms covered 9% of the ocean they could produce enough biomethane to supply Earth's equivalent demand for fossil fuel energy, remove 53 gigatonnes of CO_2 per year from the atmosphere and sustainably produce 200 kg per year of fish, per person, for 10 billion people. Ideal species for such farming and conversion include Laminaria digitata, Fucus serratus and Saccharina latissima.

Physical Processes

Biochar can be landfilled, used as a soil improver or burned using carbon capture and storage.

Biomass-related

Bio-energy with Carbon Capture and Storage (BECCS)

BECCS refers to biomass in power stations and boilers that use carbon capture and storage. The carbon sequestered by the biomass would be captured and stored, thus removing carbon dioxide from the atmosphere.

This technology is sometimes referred to as bio-energy with carbon storage, BECS, though this term can also refer to the carbon sequestration potential in other technologies, such as biochar.

Burial

Burying biomass (such as trees) directly, mimics the natural processes that created fossil fuels. Landfills also represent a physical method of sequestration.

Biochar Burial

Biochar is charcoal created by pyrolysis of biomass waste. The resulting material is added to a landfill or used as a soil improver to create terra preta. Addition of pyrogenic organic carbon (biochar) is a novel strategy to increase the soil-C stock for the long-term and to mitigate global-warming by offsetting the atmospheric C (up to 9.5 Pg C annually).

In the soil, the carbon is unavailable for oxidation to CO2 and consequential atmospheric release. This is one technique advocated by scientist James Lovelock, creator of the Gaia hypothesis. According to Simon Shackley, "people are talking more about something in the range of one to two billion tonnes a year."

The mechanisms related to biochar are referred to as bio-energy with carbon storage, BECS.

Ocean Storage

If CO_2 were to be injected to the ocean bottom, the pressures would be great enough for CO_2 to be in its liquid phase. The idea behind ocean injection would be to have stable, stationary pools of CO_2

at the ocean floor. The ocean could potentially hold over a thousand billion tons of CO_2. However, this avenue of sequestration isn't being as actively pursued because of concerns about the impact on ocean life, and concerns about its stability.

River mouths bring large quantities of nutrients and dead material from upriver into the ocean as part of the process that eventually produces fossil fuels. Transporting material such as crop waste out to sea and allowing it to sink exploits this idea to increase carbon storage. International regulations on marine dumping may restrict or prevent use of this technique.

Geological Sequestration

Geological sequestration refers to the storage of CO_2 underground in depleted oil and gas reservoirs, saline formations, or deep, un-minable coal beds.

Once CO_2 is captured from a gas or coal-fired power plant, it would be compressed to ≈100 bar so that it would be a supercritical fluid. In this fluid form, the CO_2 would be easy to transport via pipeline to the place of storage. The CO_2 would then be injected deep underground, typically around 1 km, where it would be stable for hundreds to millions of years. At these storage conditions, the density of supercritical CO_2 is 600 to 800 kg / m³. For consumers, the cost of electricity from a coal-fired power plant with carbon capture and storage (CCS) is estimated to be 0.01 - 0.05 $ / kWh higher than without CCS. For reference, the average cost of electricity in the US in 2004 was 0.0762 $ / kWh. In other terms, the cost of CCS would be 20 - 70 $/ton of CO_2 captured. The transportation and injection of CO_2 is relatively cheap, with the capture costs accounting for 70 - 80% of CCS costs.

The important parameters in determining a good site for carbon storage are: rock porosity, rock permeability, absence of faults, and geometry of rock layers. The medium in which the CO_2 is to be stored ideally has a high porosity and permeability, such as sandstone or limestone. Sandstone can have a permeability ranging from 1 to 10^{-5} Darcy, and can have a porosity as high as ≈30%. The porous rock must be capped by a layer of low permeability which acts as a seal, or caprock, for the CO_2. Shale is an example of a very good caprock, with a permeability of 10^{-5} to 10^{-9} Darcy. Once injected, the CO_2 plume will rise via buoyant forces, since it is less dense than its surroundings. Once it encounters a caprock, it will spread laterally until it encounters a gap. If there are fault planes near the injection zone, there is a possibility the CO_2 could migrate along the fault to the surface, leaking into the atmosphere, which would be potentially dangerous to life in the surrounding area. Another danger related to carbon sequestration is induced seismicity. If the injection of CO_2 creates pressures that are too high underground, the formation will fracture, causing an earthquake.

While trapped in a rock formation, CO_2 can be in the supercritical fluid phase or dissolve in groundwater/brine. It can also react with minerals in the geologic formation to precipitate carbonates.

Worldwide storage capacity in oil and gas reservoirs is estimated to be 675 - 900 Gt CO_2, and in un-minable coal seams is estimated to be 15 - 200 Gt CO_2. Deep saline formations have the largest capacity, which is estimated to be 1,000 - 10,000 Gt CO_2. In the US, there is an estimated 160 Gt CO_2 storage capacity.

There are a number of large-scale carbon capture and sequestration projects that have demonstrated the viability and safety of this method of carbon storage, which are summarized here by the Global CCS Institute. The dominant monitoring technique is seismic imaging, where vibrations are generated that propagate through the subsurface. The geologic structure can be imaged from the refracted/reflected waves.

The first large-scale CO_2 sequestration project which began in 1996 is called Sleipner, and is located in the North Sea where Norway's StatoilHydro strips carbon dioxide from natural gas with amine solvents and disposed of this carbon dioxide in a deep saline aquifer. In 2000, a coal-fueled synthetic natural gas plant in Beulah, North Dakota, became the world's first coal-using plant to capture and store carbon dioxide, at the Weyburn-Midale Carbon Dioxide Project.

CO_2 has been used extensively in enhanced crude oil recovery operations in the United States beginning in 1972. There are in excess of 10,000 wells that inject CO_2 in the state of Texas alone. The gas comes in part from anthropogenic sources, but is principally from large naturally occurring geologic formations of CO_2. It is transported to the oil-producing fields through a large network of over 5,000 kilometres (3,100 mi) of CO_2 pipelines. The use of CO_2 for enhanced oil recovery (EOR) methods in heavy oil reservoirs in the Western Canadian Sedimentary Basin (WCSB) has also been proposed. However, transport cost remains an important hurdle. An extensive CO_2 pipeline system does not yet exist in the WCSB. Athabasca oil sands mining that produces CO_2 is hundreds of kilometers north of the subsurface Heavy crude oil reservoirs that could most benefit from CO_2 injection.

Chemical Processes

Developed in the Netherlands, an electrocatalysis by a copper complex helps reduce carbon dioxide to oxalic acid; This conversion uses carbon dioxide as a feedstock to generate oxalic acid.

Mineral Carbonation

Carbon, in the form of CO_2 can be removed from the atmosphere by chemical processes, and stored in stable carbonate mineral forms. This process is known as 'carbon sequestration by mineral carbonation' or mineral sequestration. The process involves reacting carbon dioxide with abundantly available metal oxides–either magnesium oxide (MgO) or calcium oxide (CaO)–to form stable carbonates. These reactions are exothermic and occur naturally (e.g., the weathering of rock over geologic time periods).

$$CaO + CO_2 \rightarrow CaCO_3$$

$$MgO + CO_2 \rightarrow MgCO_3$$

Calcium and magnesium are found in nature typically as calcium and magnesium silicates (such as forsterite and serpentinite) and not as binary oxides. For forsterite and serpentine the reactions are:

$$Mg_2SiO_4 + 2\ CO_2 \rightarrow 2\ MgCO_3 + SiO_2$$

$$Mg3Si2O5(OH)4 + 3\ CO_2 \rightarrow 3\ MgCO_3 + 2\ SiO_2 + 2\ H_2O$$

The following table lists principal metal oxides of Earth's crust. Theoretically up to 22% of this mineral mass is able to form carbonates.

Earthen Oxide	Percent of Crust	Carbonate	Enthalpy change (kJ/mol)
SiO2	59.71		
Al2O3	15.41		
CaO	4.90	CaCO3	-179
MgO	4.36	MgCO3	-117
Na2O	3.55	Na2CO3	
FeO	3.52	FeCO3	
K2O	2.80	K2CO3	
Fe2O3	2.63	FeCO3	
	21.76	All Carbonates	

These reactions are slightly more favorable at low temperatures. This process occurs naturally over geologic time frames and is responsible for much of the Earth's surface limestone. The reaction rate can be made faster however, by reacting at higher temperatures and/or pressures, although this method requires some additional energy. Alternatively, the mineral could be milled to increase its surface area, and exposed to water and constant abrasion to remove the inert Silica as could be achieved naturally by dumping Olivine in the high energy surf of beaches Experiments suggest the weathering process is reasonably quick (one year) given porous basaltic rocks.

CO_2 naturally reacts with peridotite rock in surface exposures of ophiolites, notably in Oman. It has been suggested that this process can be enhanced to carry out natural mineralisation of CO_2.

Industrial Use

Traditional cement manufacture releases large amounts of carbon dioxide, but newly developed cement types from Novacem can absorb CO_2 from ambient air during hardening. A similar technique was pioneered by TecEco, which has been producing "EcoCement" since 2002.

In Estonia, oil shale ash, generated by power stations could be used as sorbents for CO_2 mineral sequestration. The amount of CO_2 captured averaged 60 to 65% of the carbonaceous CO_2 and 10 to 11% of the total CO_2 emissions.

Chemical Scrubbers

Various carbon dioxide scrubbing processes have been proposed to remove CO_2 from the air, usually using a variant of the Kraft process. Carbon dioxide scrubbing variants exist based on potassium carbonate, which can be used to create liquid fuels, or on sodium hydroxide. These notably include artificial trees proposed by Klaus Lackner to remove carbon dioxide from the atmosphere using chemical scrubbers.

Ocean-related

Basalt Storage

Carbon dioxide sequestration in basalt involves the injecting of CO_2 into deep-sea formations.

The CO_2 first mixes with seawater and then reacts with the basalt, both of which are alkaline-rich elements. This reaction results in the release of Ca^{2+} and Mg^{2+} ions forming stable carbonate minerals.

Underwater basalt offers a good alternative to other forms of oceanic carbon storage because it has a number of trapping measures to ensure added protection against leakage. These measures include "geothermal, sediment, gravitational and hydrate formation." Because CO_2 hydrate is denser than CO_2 in seawater, the risk of leakage is minimal. Injecting the CO_2 at depths greater than 2,700 meters (8,900 ft) ensures that the CO_2 has a greater density than seawater, causing it to sink.

One possible injection site is Juan de Fuca plate. Researchers at the Lamont-Doherty Earth Observatory found that this plate at the western coast of the United States has a possible storage capacity of 208 gigatons. This could cover the entire current U.S. carbon emissions for over 100 years.

This process is undergoing tests as part of the CarbFix project, resulting in 95% of the injected 250 tonnes of CO_2 to solidify into calcite in 2 years, using 25 tonnes of water per tonne of CO_2.

Acid Neutralisation

Carbon dioxide forms carbonic acid when dissolved in water, so ocean acidification is a significant consequence of elevated carbon dioxide levels, and limits the rate at which it can be absorbed into the ocean (the solubility pump). A variety of different bases have been suggested that could neutralize the acid and thus increase CO_2 absorption. For example, adding crushed limestone to oceans enhances the absorption of carbon dioxide. Another approach is to add sodium hydroxide to oceans which is produced by electrolysis of salt water or brine, while eliminating the waste hydrochloric acid by reaction with a volcanic silicate rock such as enstatite, effectively increasing the rate of natural weathering of these rocks to restore ocean pH.

Obstruction

Danger of Leaks

Carbon dioxide may be stored deep underground. At depth, hydrostatic pressure acts to keep it in a liquid state. Reservoir design faults, rock fissures and tectonic processes may act to release the gas stored into the ocean or atmosphere.

Financial Costs

The use of the technology would add an additional 1-5 cents of cost per kilowatt hour, according to estimate made by the Intergovernmental Panel on Climate Change. The financial costs of modern coal technology would nearly double if use of CCS technology were to be required by regulation. The cost of CCS technology differs with the different types of capture technologies being used and with the different sites that it is implemented in, but the costs tend to increase with CCS capture implementation. One study conducted predicted that with new technologies these costs could be lowered but would remain slightly higher than prices without CCS technologies.

Energy Requirements

The energy requirements of sequestration processes may be significant. In one paper, sequestration consumed 25 percent of the plant's rated 600 megawatt output capacity.

After adding CO_2 capture and compression, the capacity of the coal-fired power plant is reduced to 457 MW.

Algae Bioreactor

An algae bioreactor or photobioreactor is used for cultivating algae on purpose to fix CO_2 or produce biomass. Specifically, algae bioreactors can be used to produce fuels such as biodiesel and bioethanol, to generate animal feed, or to reduce pollutants such as NOx and CO_2 in flue gases of power plants. Fundamentally, this kind of bioreactor is based on the photosynthetic reaction which is performed by the chlorophyll-containing algae itself using dissolved carbon dioxide and sunlight energy. The carbon dioxide is dispersed into the reactor fluid to make it accessible for the algae. The bioreactor has to be made out of transparent material.

The algae are photoautotroph organisms which perform oxygenic photosynthesis.

The equation for photosynthesis:

$$6\,CO_2 + 6\,H_2O \rightarrow C_6H_{12}O_6 + 6\,O_2 \qquad \Delta H^0 = +2870\,\frac{kJ}{mol}$$

Historical Background

Some of the first experiments with the aim of cultivating algae were conducted in 1957 by the "Carnegie Institution" in Washington. In these experiments, the monocellular Chlorella were cultivated by adding CO_2 and some minerals. In the early days, bioreactors were used which were made of glass and later changed to a kind of plastic bag. The goal of all this research has been the cultivation of algae to produce a cheap animal feed.

Frequently used Photo Reactor Types

Nowadays 3 basic types of algae photobioreactors have to be differentiated, but the determining factor is the unifying parameter – the available intensity of sunlight energy.

Plate Photobioreactor

A plate reactor simply consists of vertically arranged or inclined rectangular boxes which are often divided in two parts to effect an agitation of the reactor fluid. Generally these boxes are arranged to a system by linking them. Those connections are also used for making the process of filling/emptying, introduction of gas and transport of nutritive substances, easier. The introduction of the flue gas mostly occurs at the bottom of the box to ensure that the carbon dioxide has enough time to interact with algae in the reactor fluid.

Tubular Photobioreactor

A tubular reactor consists of vertical or horizontal arranged tubes, connected together to a pipe system. The algae-suspended fluid is able to circulate in this tubing. The tubes are generally made out of transparent plastics or borosilicate glass and the constant circulation is kept up by a pump at the end of the system. The introduction of gas takes place at the end/beginning of the tube system. This way of introducing gas causes the problem of deficiency of carbon dioxide, high concentration of oxygen at the end of the unit during the circulation, and bad efficiency.

Bubble Column Photobioreactor

A bubble column photo reactor consists of vertical arranged cylindrical column, made out of transparent material. The introduction of gas takes place at the bottom of the column and causes a turbulent stream to enable an optimum gas exchange. At present these types of reactors are built with a maximum diameter of 20 cm to 30 cm in order to ensure the required supply of sunlight energy.

The biggest problem with the sunlight determined construction is the limited size of the diameter. Feuermann et al. invented a method to collect sunlight with a cone shaped collector and transfer it with some fiberglass cables which are adapted to the reactor in order to enable constructions of a column reactor with wider diameters. - on this scale the energy consumption due to pumps etc. and the CO_2 cost of manufacture may outweigh the CO_2 captured by the reactor.

Industrial Usage

The cultivation of algae in a photobioreactor creates a narrow range of industrial application possibilities. Some power companies already established research facilities with algae photobioreactors to find out how efficient they could be in reducing CO_2 emissions, which are contained in flue gas, and how much biomass will be produced. Algae biomass has many uses and can be sold to generate additional income. The saved emission volume can bring an income too, by selling emission credits to other power companies.

The utilisation of algae as food is very common in East Asian regions. Most of the species contain only a fraction of usable proteins and carbohydrates, and a lot of minerals and trace elements. Generally, the consumption of algae should be minimal because of the high iodine content. For example, if someone has hyperthyroidism it could be very dangerous for their health. Likewise, many species of diatomaceous algae produce compounds unsafe for humans.

The algae, especially some species which contain over 50 percent oil and a lot of carbohydrates, can be used for producing biodiesel and bioethanol by extracting and refining the fractions. This point is very interesting, because the algae biomass is generated 30 times faster than some agricultural biomass, which is commonly used for producing biodiesel.

Climate Change Mitigation

Climate change mitigation consists of actions to limit the magnitude or rate of long-term climate change. Climate change mitigation generally involves reductions in human (anthropogenic) emissions of greenhouse gases (GHGs). Mitigation may also be achieved by increasing the capacity of carbon sinks, e.g., through reforestation. Mitigation policies can substantially reduce the risks associated with human-induced global warming.

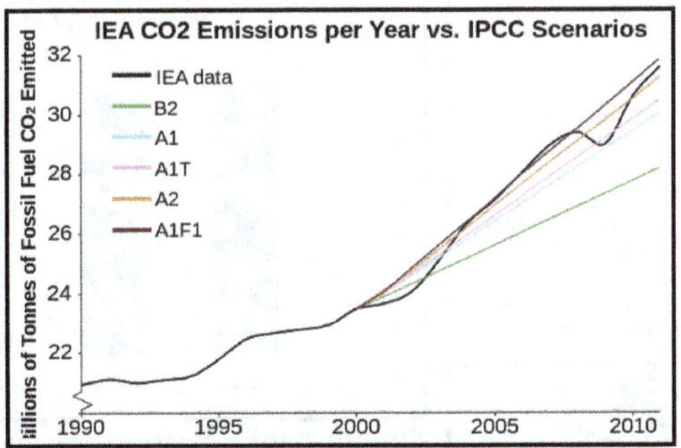

Fossil fuel related CO_2 emissions compared to five of IPCC's emissions scenarios. The dips are related to global recessions. Data from IPCC SRES scenarios; Data spreadsheet included with International Energy Agency's "CO_2 Emissions from Fuel Combustion 2010 – Highlights"; and Supplemental IEA data.

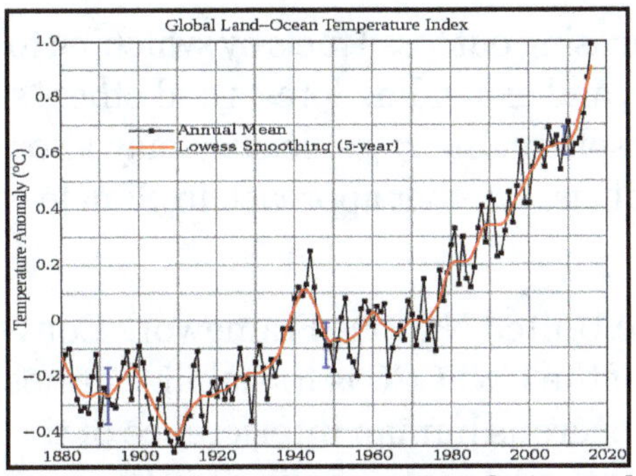

Global mean surface temperature change from 1880 to 2016, relative to the 1951–1980 mean. The black line is the global annual mean and the red line is the five-year lowess smooth. The blue uncertainty bars show a 95% confidence limit. Global dimming, from sulfate aerosol air pollution, between 1950 and 1980 is believed to have mitigated global warming somewhat.

According to the IPCC's 2014 assessment report, "Mitigation is a public good; climate change is a case of the 'tragedy of the commons'. Effective climate change mitigation will not be achieved if each agent (individual, institution or country) acts independently in its own selfish interest, suggesting the need for collective action. Some adaptation actions, on the other hand, have characteristics of a private good as benefits of actions may accrue more directly to the individuals, regions, or countries that undertake them, at least in the short term. Nevertheless, financing such adaptive activities remains an issue, particularly for poor individuals and countries."

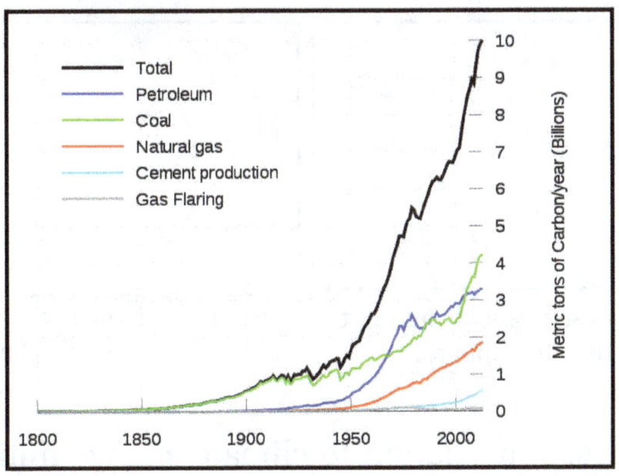

Global carbon dioxide emissions from human activities 1800–2007.

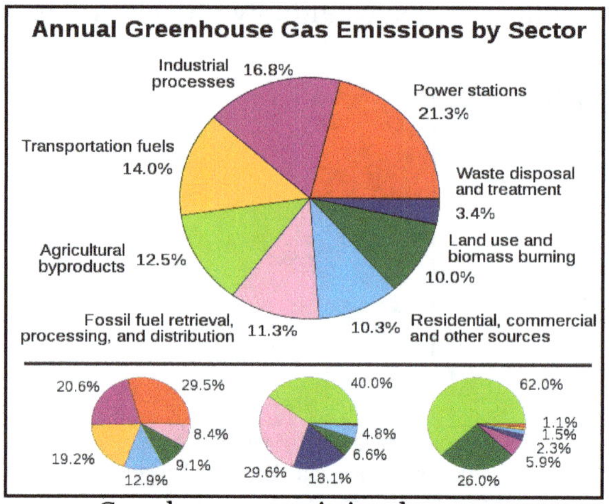

Greenhouse gas emissions by sector.

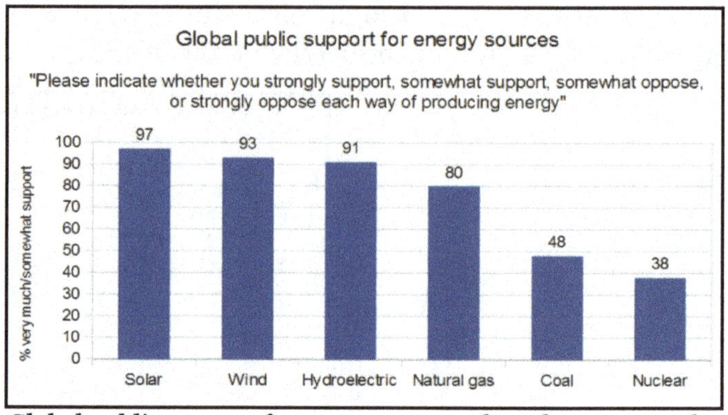

Global public support for energy sources, based on a survey by Ipsos (2011).

Examples of mitigation include phasing out fossil fuels by switching to low-carbon energy sources, such as renewable and nuclear energy, and expanding forests and other "sinks" to remove greater amounts of carbon dioxide from the atmosphere. Energy efficiency may also play a role, for example, through improving the insulation of buildings. Another approach to climate change mitigation is climate engineering.

Most countries are parties to the United Nations Framework Convention on Climate Change (UNFCCC). The ultimate objective of the UNFCCC is to stabilize atmospheric concentrations of GHGs at a level that would prevent dangerous human interference of the climate system. Scientific analysis can provide information on the impacts of climate change, but deciding which impacts are dangerous requires value judgments.

In 2010, Parties to the UNFCCC agreed that future global warming should be limited to below 2.0 °C (3.6 °F) relative to the pre-industrial level. With the Paris Agreement of 2015 this was confirmed, but was revised with a new target laying down "parties will do the best" to achieve warming below 1.5 °C. The current trajectory of global greenhouse gas emissions does not appear to be consistent with limiting global warming to below 1.5 or 2 °C. Other mitigation policies have been proposed, some of which are more stringent or modest than the 2 °C limit.

Greenhouse Gas Concentrations and Stabilization

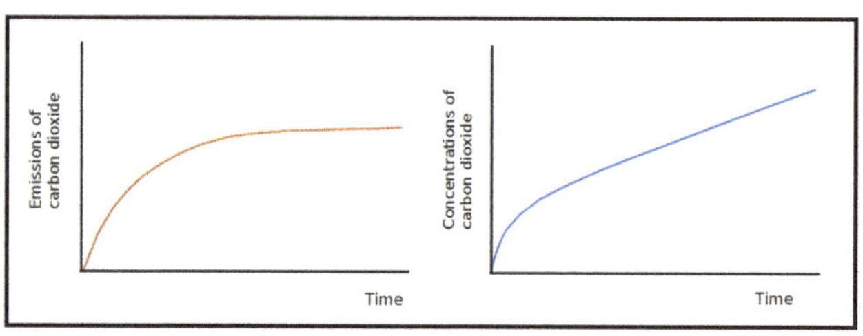

Stabilizing CO_2 emissions at their present level would not stabilize its concentration in the atmosphere.

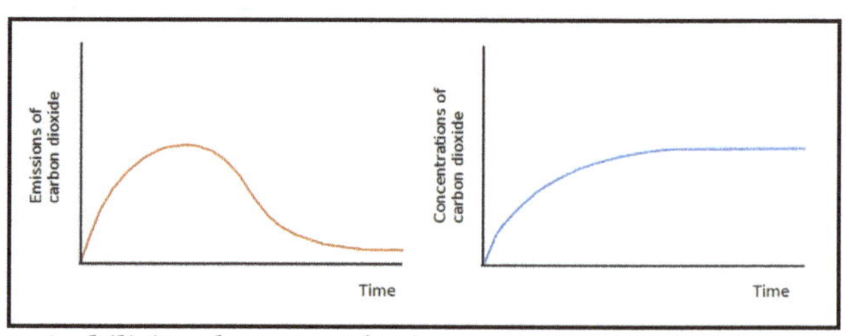

Stabilizing the atmospheric concentration of CO_2 at a constant level would require emissions to be effectively eliminated.

One of the issues often discussed in relation to climate change mitigation is the stabilization of greenhouse gas concentrations in the atmosphere. The United Nations Framework Convention on

Climate Change (UNFCCC) has the ultimate objective of preventing "dangerous" anthropogenic (i.e., human) interference of the climate system. As is stated in Article 2 of the Convention, this requires that greenhouse gas (GHG) concentrations are stabilized in the atmosphere at a level where ecosystems can adapt naturally to climate change, food production is not threatened, and economic development can proceed in a sustainable fashion.

There are a number of anthropogenic greenhouse gases. These include carbon dioxide (chemical formula: CO_2), methane (CH_4), nitrous oxide (N_2O), and a group of gases referred to as halocarbons. The emissions reductions necessary to stabilize the atmospheric concentrations of these gases varies. CO_2 is the most important of the anthropogenic greenhouse gases.

There is a difference between stabilizing CO_2 emissions and stabilizing atmospheric concentrations of CO_2. Stabilizing emissions of CO_2 at current levels would not lead to a stabilization in the atmospheric concentration of CO_2. In fact, stabilizing emissions at current levels would result in the atmospheric concentration of CO_2 continuing to rise over the 21st century and beyond.

The reason for this is that human activities are adding CO_2 to the atmosphere faster than natural processes can remove it. This is analogous to a flow of water into a bathtub. So long as the tap runs water (analogous to the emission of carbon dioxide) into the tub faster than water escapes through the plughole (the natural removal of carbon dioxide from the atmosphere), then the level of water in the tub (analogous to the concentration of carbon dioxide in the atmosphere) will continue to rise.

According to some studies, stabilizing atmospheric CO_2 concentrations would require anthropogenic CO_2 emissions to be reduced by 80% relative to the peak emissions level. An 80% reduction in emissions would stabilize CO_2 concentrations for around a century, but even greater reductions would be required beyond this. Other research has found that, after leaving room for emissions for food production for 9 billion people and to keep the global temperature rise below 2 °C, emissions from energy production and transport will have to peak almost immediately in the developed world and decline at ca. 10% per annum until zero emissions are reached around 2030. In developing countries energy and transport emissions would have to peak by 2025 and then decline similarly.

Stabilizing the atmospheric concentration of the other greenhouse gasses humans emit also depends on how fast their emissions are added to the atmosphere, and how fast the GHGs are removed. Stabilization for these gases is described in the later section on non-CO_2 GHGs.

Projections

Projections of future greenhouse gas emissions are highly uncertain. In the absence of policies to mitigate climate change, GHG emissions could rise significantly over the 21st century.

Numerous assessments have considered how atmospheric GHG concentrations could be stabilized. The lower the desired stabilization level, the sooner global GHG emissions must peak and decline. GHG concentrations are unlikely to stabilize this century without major policy changes.

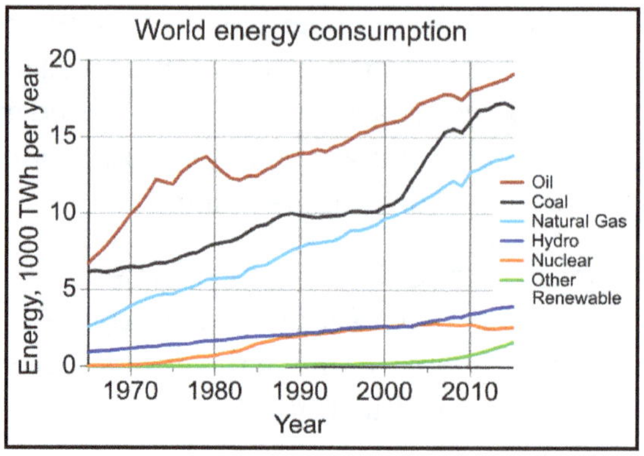

The above image shows rate of world energy usage per day, from 1970 to 2010. Every fossil fuel source has increased in large amounts between 1970 and 2010, dominating all other energy sources. Hydroelectricity has increased at a slow steady rate over this same period, nuclear entered a period of rapid growth between 1970 and 1990 before leveling off. Other renewables, between 2000 and 2010 have, having started from a low usage rate, began to enter into a period of rapid growth. 1000 TWh=1 PWh.

Energy Consumption by Power Source

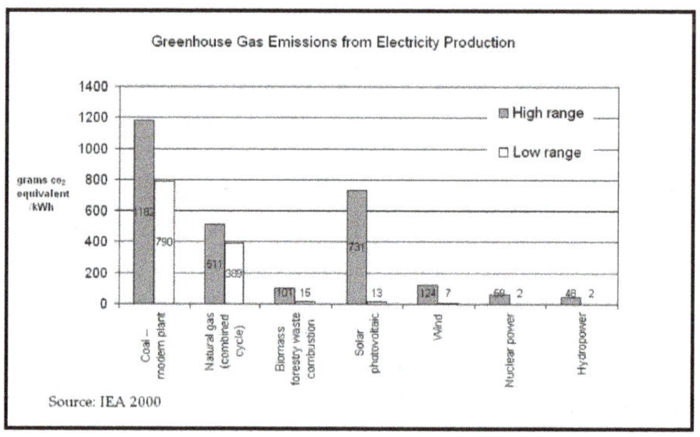

"Hydropower-Internalised Costs and Externalised Benefits"; Frans H. Koch; International Energy Agency (IEA)-Implementing Agreement for Hydropower Technologies and Programmes; 2000.

To create lasting climate change mitigation, the replacement of high carbon emission intensity power sources, such as conventional fossil fuels—oil, coal and natural gas—with low-carbon power sources is required. Fossil fuels supply humanity with the vast majority of our energy demands, and at a growing rate. In 2012 the IEA noted that coal accounted for half the increased energy use of the prior decade, growing faster than all renewable energy sources. Both hydroelectricity and nuclear power together provide the majority of the generated low-carbon power fraction of global total power consumption.

Fuel type	Average total global power consumption in TW		
	1980	2004	2006
Oil	4.38	5.58	5.74
Gas	1.80	3.45	3.61
Coal	2.34	3.87	4.27

Hydroelectric	0.60	0.93	1.00
Nuclear power	0.25	0.91	0.93
Geothermal, wind, solar energy, wood	0.02	0.13	0.16
Total	**9.48**	**15.0**	**15.8**

Change and use of energy, by source, in units of (PWh) in that year.				
	Fossil	**Nuclear**	**All renewables**	**Total**
1990	83.374	6.113	13.082	102.569
2000	94.493	7.857	15.337	117.687
2008	117.076	8.283	18.492	143.851
Change 2000–2008	22.583	0.426	3.155	26.164

Methods and Means

Assessments often suggest that GHG emissions can be reduced using a portfolio of low-carbon technologies. At the core of most proposals is the reduction of greenhouse gas (GHG) emissions through reducing energy waste and switching to low-carbon power sources of energy. As the cost of reducing GHG emissions in the electricity sector appears to be lower than in other sectors, such as in the transportation sector, the electricity sector may deliver the largest proportional carbon reductions under an economically efficient climate policy.

"Economic tools can be useful in designing climate change mitigation policies." "While the limitations of economics and social welfare analysis, including cost–benefit analysis, are widely documented, economics nevertheless provides useful tools for assessing the pros and cons of taking, or not taking, action on climate change mitigation, as well as of adaptation measures, in achieving competing societal goals. Understanding these pros and cons can help in making policy decisions on climate change mitigation and can influence the actions taken by countries, institutions and individuals."

Other frequently discussed means include energy conservation, increasing fuel economy in automobiles (which includes the use of electric hybrids), charging plug-in hybrids and electric cars by low-carbon electricity, making individual-lifestyle changes (e.g., cycling instead of driving), and changing business practices. Many fossil fuel driven vehicles can be converted to use electricity, the US has the potential to supply electricity for 73% of light duty vehicles (LDV), using overnight charging. The US average CO_2 emissions for a battery-electric car is 180 grams per mile vs 430 grams per mile for a gasoline car. The emissions would be displaced away from street level, where they have "high human-health implications. Increased use of electricity "generation for meeting the future transportation load is primarily fossil-fuel based", mostly natural gas, followed by coal, but could also be met through nuclear, tidal, hydroelectric and other sources.

A range of energy technologies may contribute to climate change mitigation. These include nuclear power and renewable energy sources such as biomass, hydroelectricity, wind power, solar power, geothermal power, ocean energy, and; the use of carbon sinks, and carbon capture and storage. For example, Pacala and Socolow of Princeton have proposed a 15 part program to reduce CO_2 emissions by 1 billion metric tons per year – or 25 billion tons over the 50-year period using today's technologies as a type of Global warming game.

Another consideration is how future socio-economic development proceeds. Development choices (or "pathways") can lead differences in GHG emissions. Political and social attitudes may affect how easy or difficult it is to implement effective policies to reduce emissions.

Alternative Energy Sources

Renewable Energy

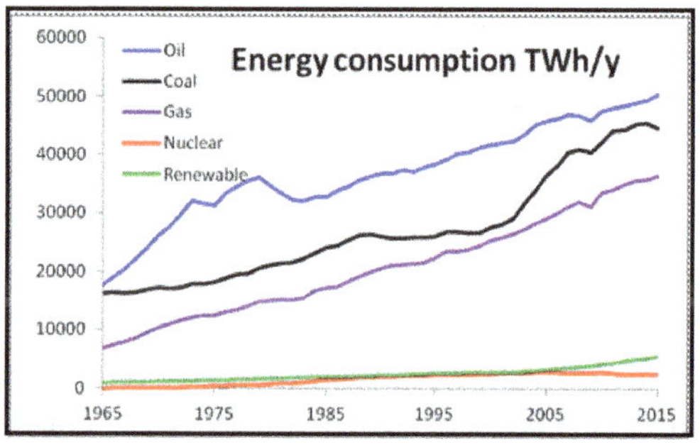

The worldwide growth of renewable energy is shown by the green line.

The 22,500 MW nameplate capacity Three Gorges Dam in the Peoples Republic of China, the largest hydroelectric power station in the world.

The Shepherds Flat Wind Farm is an 845 megawatt (MW) nameplate capacity, wind farm in the US state of Oregon, each turbine is a nameplate 2 or 2.5 MW electricity generator.

The 150 MW Andasol solar power station is a commercial parabolic trough solar thermal power plant, located in Spain. The Andasol plant uses tanks of molten salt to store solar energy so that it can continue generating electricity for 7.5 hours after the sun has stopped shining.

Renewable energy flows involve natural phenomena such as sunlight, wind, rain, tides, plant growth, and geothermal heat, as the International Energy Agency explains:

Renewable energy is derived from natural processes that are replenished constantly. In its various forms, it derives directly from the sun, or from heat generated deep within the earth. Included in the definition is electricity and heat generated from solar, wind, ocean, hydropower, biomass, geothermal resources, and biofuels and hydrogen derived from renewable resources.

Climate change concerns and the need to reduce carbon emissions are driving increasing growth in the renewable energy industries. Low-carbon renewable energy replaces conventional fossil fuels in three main areas: power generation, hot water/ space heating, and transport fuels. In 2011, the share of renewables in electricity generation worldwide grew for the fourth year in a row to 20.2%. Based on REN21's 2014 report, renewables contributed 19% to supply global energy consumption. This energy consumption is divided as 9% coming from burning biomass, 4.2% as heat energy (non-biomass), 3.8% hydro electricity and 2% as electricity from wind, solar, geothermal, and biomass thermal power plants.

Renewable energy use has grown much faster than anyone anticipated. The Intergovernmental Panel on Climate Change (IPCC) has said that there are few fundamental technological limits to integrating a portfolio of renewable energy technologies to meet most of total global energy demand. At the national level, at least 30 nations around the world already have renewable energy contributing more than 20% of energy supply.

As of 2012, renewable energy accounts for almost half of new electricity capacity installed and costs are continuing to fall. Public policy and political leadership helps to "level the playing field" and drive the wider acceptance of renewable energy technologies. As of 2011, 118 countries have targets for their own renewable energy futures, and have enacted wide-ranging public policies to promote renewables. Leading renewable energy companies include BrightSource Energy, First Solar, Gamesa, GE Energy, Goldwind, Sinovel, Suntech, Trina Solar, Vestas and Yingli.

The incentive to use 100% renewable energy has been created by global warming and other ecological as well as economic concerns. Mark Z. Jacobson says producing all new energy with wind power, solar power, and hydropower by 2030 is feasible and existing energy supply arrangements could be replaced by 2050. Barriers to implementing the renewable energy plan are seen to be "primarily social and political, not technological or economic". Jacobson says that energy costs with a wind, solar, water system should be similar to today's energy costs. According to a 2011 projection by the (IEA)International Energy Agency, solar power generators may produce most of the world's electricity within 50 years, dramatically reducing harmful greenhouse gas emissions. Critics of the "100% renewable energy" approach include Vaclav Smil and James E. Hansen. Smil and Hansen are concerned about the variable output of solar and wind power, NIMBYism, and a lack of infrastructure.

Economic analysts expect market gains for renewable energy (and efficient energy use) following the 2011 Japanese nuclear accidents. In his 2012 State of the Union address, President Barack Obama restated his commitment to renewable energy and mentioned the long-standing Interior Department commitment to permit 10,000 MW of renewable energy projects on public land in 2012. Globally, there are an estimated 3 million direct jobs in renewable energy industries, with about half of them in the biofuels industry.

Some countries, with favorable geography, geology and weather well suited to an economical exploitation of renewable energy sources, already get most of their electricity from renewables, including from geothermal energy in Iceland (100 percent), and Hydroelectric power in Brazil (85 percent), Austria (62 percent), New Zealand (65 percent), and Sweden (54 percent). Renewable power generators are spread across many countries, with wind power providing a significant share of electricity in some regional areas: for example, 14 percent in the US state of Iowa, 40 percent in the northern German state of Schleswig-Holstein, and 20 percent in Denmark. Solar water heating makes an important and growing contribution in many countries, most notably in China, which now has 70 percent of the global total (180 GWth). Worldwide, total installed solar water heating systems meet a portion of the water heating needs of over 70 million households. The use of biomass for heating continues to grow as well. In Sweden, national use of biomass energy has surpassed that of oil. Direct geothermal heating is also growing rapidly. Renewable biofuels for transportation, such as ethanol fuel and biodiesel, have contributed to a significant decline in oil consumption in the United States since 2006. The 93 billion liters of biofuels produced worldwide in 2009 displaced the equivalent of an estimated 68 billion liters of gasoline, equal to about 5 percent of world gasoline production.

Some of the world's largest solar power stations: Ivanpah (CSP) and Topaz (PV), both in California.

Nuclear Power

Since about 2001 the term "nuclear renaissance" has been used to refer to a possible nuclear power industry revival, driven by rising fossil fuel prices and new concerns about meeting greenhouse gas emission limits. However, in March 2011 the Fukushima nuclear disaster in Japan and associated shutdowns at other nuclear facilities raised questions among some commentators over the future of nuclear power. Platts has reported that "the crisis at Japan's Fukushima nuclear plants has prompted leading energy-consuming countries to review the safety of their existing reactors and cast doubt on the speed and scale of planned expansions around the world".

The World Nuclear Association has reported that nuclear electricity generation in 2012 was at its lowest level since 1999. Several previous international studies and assessments, suggested that as part of the portfolio of other low-carbon energy technologies, nuclear power will continue to play a role in reducing greenhouse gas emissions. Historically, nuclear power usage is estimated to have prevented the atmospheric emission of 64 gigatonnes of CO_2-equivalent as of 2013. Public concerns about nuclear power include the fate of spent nuclear fuel, nuclear accidents, security risks, nuclear proliferation, and a concern that nuclear power plants are very expensive. Of these concerns, nuclear accidents and disposal of long-lived radioactive fuel/"waste" have probably had the greatest public impact worldwide. Although generally unaware of it, both of these glaring public concerns are greatly diminished by present passive safety designs, the experimentally proven, "melt-down proof" EBR-II, future molten salt reactors, and the use of conventional and more advanced fuel/"waste" pyroprocessing, with the latter recycling or reprocessing not presently being commonplace as it is often considered to be cheaper to use a once-through nuclear fuel cycle in many countries, depending on the varying levels of intrinsic value given by a society in reducing the long-lived waste in their country, with France doing a considerable amount of reprocessing when compared to the US.

Blue Cherenkov light being produced near the core of the Fission powered Advanced Test Reactor.

Nuclear power, with a 10.6% share of world electricity production as of 2013, is second only to hydroelectricity as the largest source of low-carbon power. Over 400 reactors generate electricity in 31 countries.

A Yale University review published in the Journal of Industrial Ecology analyzing CO_2 life cycle assessment(LCA) emissions from nuclear power(Light water reactors) determined that: "The collective LCA literature indicates that life cycle GHG emissions from nuclear power are only a fraction of traditional fossil sources and comparable to renewable technologies." While some have raised uncertainty surrounding the future GHG emissions of nuclear power as a result of an extreme potential decline in uranium ore grade without a corresponding increase in the efficiency of enrichment methods. In a scenario analysis of future global nuclear development, as it could be effected by a decreasing global uranium market of average ore grade, the analysis determined that depending on conditions, median life cycle nuclear power GHG emissions could be between 9 and 110 g CO_2-eq/kWh by 2050, with the latter high figure being derived from a "worst-case scenario" that is not "considered very robust" by the authors of the paper, as the "ore grade" in the scenario is lower than the uranium concentration in many lignite coal ashes.

Although this future analyses primarily deals with extrapolations for present Generation II reactor technology, the same paper also summarizes the literature on "FBRs"/Fast Breeder Reactors, of which two are in operation as of 2014 with the newest being the BN-800, for these reactors it states that the "median life cycle GHG emissions ... [are] similar to or lower than [present light water reactors] LWRs and purports to consume little or no uranium ore.

In their 2014 report, the IPCC comparison of energy sources global warming potential per unit of electricity generated, which notably included albedo effects, mirror the median emission value derived from the Warner and Heath Yale meta-analysis for the more common non-breeding Light water reactors, a CO2-equivalent value of 12 g CO_2-eq/kWh, which is the lowest global warming forcing of all baseload power sources, with comparable low carbon power baseload sources, such as hydropower and biomass, producing substantially more global warming forcing 24 and 230 g CO_2-eq/kWh respectively.

In 2014, Brookings Institution published *The Net Benefits of Low and No-Carbon Electricity Technologies* which states, after performing an energy and emissions cost analysis, that "The net benefits of new nuclear, hydro, and natural gas combined cycle plants far outweigh the net benefits of new wind or solar plants", with the most cost effective low carbon power technology being determined to be nuclear power.

During his presidential campaign, Barack Obama stated, "Nuclear power represents more than 70% of our noncarbon generated electricity. It is unlikely that we can meet our aggressive climate goals if we eliminate nuclear power as an option."

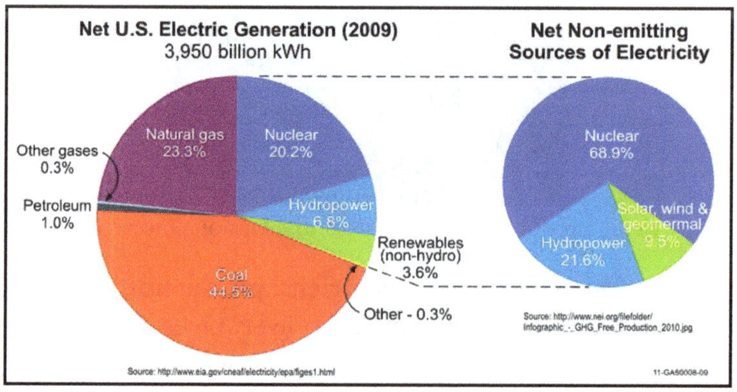

This graph illustrates nuclear power is the United States's largest contributor of non-greenhouse-gas-emitting electric power generation, comprising nearly three-quarters of the non-emitting sources.

Analysis in 2015 by Professor and Chair of Environmental Sustainability Barry W. Brook and his colleagues on the topic of replacing fossil fuels entirely, from the electric grid of the world, has determined that at the historically modest and proven-rate at which nuclear energy was added to and replaced fossil fuels in France and Sweden during each nation's building programs in the 1980s, within 10 years nuclear energy could displace or remove fossil fuels from the electric grid completely, "allow[ing] the world to meet the most stringent greenhouse-gas mitigation targets.". In a similar analysis, Brook had earlier determined that 50% of all global energy, that is not solely electricity, but transportation synfuels etc. could be generated within approximately 30 years, if the global nuclear fission build rate was identical to each of these nation's already proven decadal rates(in units of installed nameplate capacity, GW per year, per unit of global GDP(GW/year/$).

This is in contrast to the completely conceptual paper-studies for a *100% renewable energy* world, which would require an orders of magnitude more costly global investment per year, an investment rate that has no historical precedent, having never been attempted due to its prohibitive cost, and with far greater land area that would be required to be devoted to the wind, wave and solar projects, along with the inherent assumption that humanity will use less, and not more, energy in the future. As Brook notes the "principal limitations on nuclear fission are not technical, economic or fuel-related, but are instead linked to complex issues of societal acceptance, fiscal and political inertia, and inadequate critical evaluation of the real-world constraints facing [the other] low-carbon alternatives."

Nuclear power may be uncompetitive compared with fossil fuel energy sources in countries without a carbon tax program, and in comparison to a fossil fuel plant of the same power output, nuclear power plants take a longer amount of time to construct.

Two new, first of their kind, EPR reactors under construction in Finland and France have been delayed and are running over-budget. However learning from experience, two further EPR reactors under construction in China are on, and ahead, of schedule respectively. As of 2013, according to the IAEA and the European Nuclear Society, worldwide there were 68 civil nuclear power reactors under construction in 15 countries. China has 29 of these nuclear power reactors under construction, as of 2013, with plans to build many more, while in the US the licenses of almost half its reactors have been extended to 60 years, and plans to build another dozen are under serious consideration. There are also a considerable number of new reactors being built in South Korea, India, and Russia. At least 100 older and smaller reactors will "most probably be closed over the next 10–15 years". This is probable only if one does not factor in the ongoing Light Water Reactor Sustainability Program, created to permit the extension of the life span of the USA's 104 nuclear reactors to 60 years. The licenses of almost half of the USA's reactors have been extended to 60 years as of 2008. Two new "passive safety" AP1000 reactors are, as of 2013, being constructed at Vogtle Electric Generating Plant.

Public opinion about nuclear power varies widely between countries. A poll by Gallup International (2011) assessed public opinion in 47 countries. The poll was conducted following a tsunami and earthquake which caused an accident at the Fukushima nuclear power plant in Japan. 49% stated that they held favourable views about nuclear energy, while 43% held an unfavourable view. Another global survey by Ipsos (2011) assessed public opinion in 24 countries. Respondents to this survey showed a clear preference for renewable energy sources over coal and nuclear energy (refer to graph opposite). Ipsos (2012) found that solar and wind were viewed by the public as being more environmentally friendly and more viable long-term energy sources relative to nuclear power and natural gas. However, solar and wind were viewed as being less reliable relative to nuclear power and natural gas. In 2012 a poll done in the UK found that 63% of those surveyed support nuclear power, and with opposition to nuclear power at 11%. In Germany, strong anti-nuclear sentiment led to eight of the seventeen operating reactors being permanently shut down following the March 2011 Fukushima nuclear disaster.

Nuclear fusion research, in the form of the International Thermonuclear Experimental Reactor is underway. Fusion powered electricity generation was initially believed to be readily achievable, as fission power had been. However, the extreme requirements for continuous reactions and plasma containment led to projections being extended by several decades. In 2010, more than 60 years after the first attempts, commercial power production was still believed to be unlikely before 2050. Although rather than an either, or, issue economical fusion-fission hybrid reactors could be built before any attempt at this more demanding commercial "pure-fusion reactor"/DEMO reactor takes place.

Coal to Gas Fuel Switching

Most mitigation proposals imply—rather than directly state—an eventual reduction in global fossil fuel production. Also proposed are direct quotas on global fossil fuel production.

Natural gas emits far fewer greenhouse gases (i.e. CO_2 and methane—CH_4) than coal when burned at power plants, but evidence has been emerging that this benefit could be completely negated by methane leakage at gas drilling fields and other points in the supply chain.

A study performed by the Environmental Protection Agency (EPA) and the Gas Research Institute (GRI) in 1997 sought to discover whether the reduction in carbon dioxide emissions from increased natural gas (predominantly methane) use would be offset by a possible increased level of methane emissions from sources such as leaks and emissions. The study concluded that the reduction in emissions from increased natural gas use outweighs the detrimental effects of increased methane emissions. More recent peer-reviewed studies have challenged the findings of this study, with researchers from the National Oceanic and Atmospheric Administration (NOAA) reconfirming findings of high rates of methane (CH_4) leakage from natural gas fields.

A 2011 study by noted climate research scientist, Tom Wigley, found that while carbon dioxide (CO_2) emissions from fossil fuel combustion may be reduced by using natural gas rather than coal to produce energy, it also found that additional methane (CH_4) from leakage adds to the radiative forcing of the climate system, offsetting the reduction in CO_2 forcing that accompanies the transition from coal to gas. The study looked at methane leakage from coal mining; changes in radiative forcing due to changes in the emissions of sulfur dioxide and carbonaceous aerosols; and differences in the efficiency of electricity production between coal- and gas-fired power generation. On balance, these factors more than offset the reduction in warming due to reduced CO_2 emissions. When gas replaces coal there is additional warming out to 2,050 with an assumed leakage rate of 0%, and out to 2,140 if the leakage rate is as high as 10%. The overall effects on global-mean temperature over the 21st century, however, are small. Petron et al. (2013) and Alvarez et al. (2012) note that estimated that leakage from gas infrastructure is likely to be underestimated. These studies indicate that the exploitation of natural gas as a "cleaner" fuel is questionable. A 2014 meta-study of 20 years of natural gas technical literature shows that methane emissions are consistently underestimated but on a 100-year scale, the climate benefits of coal to gas fuel switching are likely larger than the negative effects of natural gas leakage.

Heat Pump

A heat pump is a device that provides heat energy from a source of heat to a destination called a "heat sink". Heat pumps are designed to move thermal energy opposite to the direction of spontaneous heat flow by absorbing heat from a cold space and releasing it to a warmer one. A heat pump uses some amount of external power to accomplish the work of transferring energy from the heat source to the heat sink.

While air conditioners and freezers are familiar examples of heat pumps, the term "heat pump" is more general and applies to many HVAC (heating, ventilating, and air conditioning) devices used for space heating or space cooling. When a heat pump is used for heating, it employs the same basic refrigeration-type cycle used by an air conditioner or a refrigerator, but in the opposite direction—releasing heat into the conditioned space rather than the surrounding environment. In this use, heat pumps generally draw heat from the cooler external air or from the ground. In heating mode, heat pumps are three to four times more efficient in their use of electric power than simple electrical resistance heaters.

Outside unit of an air-source heat pump.

It has been concluded that heat pumps are the single technology that could reduce the greenhouse gas emissions of households better than every other technology that is available on the market. With a market share of 30% and (potentially) clean electricity, heat pumps could reduce global CO_2 emissions by 8% annually. Using ground source heat pumps could reduce around 60% of the primary energy demand and 90% of CO_2 emissions in Europe in 2050 and make handling high shares of renewable energy easier. Using surplus renewable energy in heat pumps is regarded as the most effective household means to reduce global warming and fossil fuel depletion.

With significant amounts of fossil fuel used in electricity production, demands on the electrical grid also generate greenhouse gases. Without a high share of low-carbon electricity, a domestic heat pump will produce more carbon emissions than using natural gas.

Fossil Fuel Phase-out: Carbon Neutral and Negative Fuels

3,500–4,000 environmental activists blocking a coal mine in Germany to limit climate change (Ende Gelände 2016).

Fossil fuel may be phased-out with carbon neutral and carbon negative pipeline and transportation fuels created with power to gas and gas to liquids technologies. Carbon dioxide from fossil fuel flue gas can be used to produce plastic lumber allowing carbon negative reforestation.

Demand Side Management

Energy Efficiency and Conservation

A spiral-type integrated compact fluorescent lamp, use has grown among North American consumers since its introduction in the mid-1990s.

Efficient energy use, sometimes simply called "energy efficiency", is the goal of efforts to reduce the amount of energy required to provide products and services. For example, insulating a home allows a building to use less heating and cooling energy to achieve and maintain a comfortable temperature. Installing fluorescent lights or natural skylights reduces the amount of energy required to attain the same level of illumination compared to using traditional incandescent light bulbs. Compact fluorescent lights use two-thirds less energy and may last 6 to 10 times longer than incandescent lights.

Energy efficiency has proved to be a cost-effective strategy for building economies without necessarily growing energy consumption. For example, the state of California began implementing energy-efficiency measures in the mid-1970s, including building code and appliance standards with strict efficiency requirements. During the following years, California's energy consumption has remained approximately flat on a per capita basis while national US consumption doubled. As part of its strategy, California implemented a "loading order" for new energy resources that puts energy efficiency first, renewable electricity supplies second, and new fossil-fired power plants last.

Energy conservation is broader than energy efficiency in that it encompasses using less energy to achieve a lesser energy demanding service, for example through behavioral change, as well as encompassing energy efficiency. Examples of conservation without efficiency improvements would be heating a room less in winter, driving less, or working in a less brightly lit room. As with other definitions, the boundary between efficient energy use and energy conservation can be fuzzy, but both are important in environmental and economic terms. This is especially the case when actions are directed at the saving of fossil fuels.

Reducing energy use is seen as a key solution to the problem of reducing greenhouse gas emissions. According to the International Energy Agency, improved energy efficiency in buildings, industrial processes and transportation could reduce the world's energy needs in 2050 by one third, and help control global emissions of greenhouse gases.

Demand Side Switching Sources

Fuel switching on the demand side refers to changing the type of fuel used to satisfy a need for an energy service. To meet deep decarbonization goals, like the 80% reduction by 2050 goal being discussed in California and the European Union, many primary energy changes are needed. Energy efficiency alone may not be sufficient to meet these goals, switching fuels used on the demand side will help lower carbon emissions. Progressively coal, oil and eventually natural gas for space and water heating in buildings will need to be reduced. For an equivalent amount of heat, burning natural gas produces about 45 per cent less carbon dioxide than burning coal. There are various ways in which this could happen, and different strategies will likely make sense in different locations. While the system efficiency of a gas furnace may be higher than the combination of natural gas power plant and electric heat, the combination of the same natural gas power plant and an electric heat pump has lower emissions per unit of heat delivered in all but the coldest climates. This is possible because of the very efficient coefficient of performance of heat pumps.

At the beginning of this century 70% of all electricity was generated by fossil fuels, and as carbon free sources eventually make up half of the generation mix, replacing gas or oil furnaces and water heaters with electric ones will have a climate benefit. In areas like Norway, Brazil and Quebec that have abundant hydroelectricity, electric heat and hot water is common.

The economics of switching the demand side from fossil fuels to electricity for heating, will depend on the price of fuels vs electricity and the relative prices of the equipment. The EIA Annual Energy Outlook 2014 suggests that domestic gas prices will rise faster than electricity prices which will encourage electrification in the coming decades. Electrifying heating loads may also provide a flexible resource that can participate in demand response. Since thermostatically controlled loads have inherent energy storage, electrification of heating could provide a valuable resource to integrate variable renewable resources into the grid.

Alternatives to electrification, include decarbonizing pipeline gas through power to gas, biogas, or other carbon neutral fuels. A 2015 study by Energy+Environmental Economics shows that a hybrid approach of decarbonizing pipeline gas, electrification, and energy efficiency can meet carbon reduction goals at a similar cost as only electrification and energy efficiency in Southern California.

Demand Side Grid Management

Expanding intermittent electrical sources such as wind power, creates a growing problem balancing grid fluctuations. Some of the plans include building pumped storage or continental super grids costing billions of dollars. However instead of building for more power,there are a variety of ways to affect the size and timing of electricity demand on the consumer side. Designing for reduced demands on a smaller power grid is more efficient and economic than having extra generation and transmission for intermittentcy, power failures and peak demands. Having these abilities is one of the chief aims of a smart grid.

Time of use metering is a common way to motivate electricity users to reduce their peak load consumption. For instance, running dishwashers and laundry at night after the peak has passed, reduces electricity costs.

Dynamic demand plans have devices passively shut off when stress is sensed on the electrical grid.

This method may work very well with thermostats, when power on the grid sags a small amount, a low power temperature setting is automatically selected reducing the load on the grid. For instance millions of refrigerators reduce their consumption when clouds pass over solar installations. Consumers would need to have a smart meter in order for the utility to calculate credits.

Demand response devices could receive all sorts of messages from the grid. The message could be a request to use a low power mode similar to dynamic demand, to shut off entirely during a sudden failure on the grid, or notifications about the current and expected prices for power. This would allow electric cars to recharge at the least expensive rates independent of the time of day. The vehicle-to-grid suggestion would use a car's battery or fuel cell to supply the grid temporarily.

Lifestyle and Behavior

The IPCC Fifth Assessment Report emphasises that behaviour, lifestyle and cultural change have a high mitigation potential in some sectors, particularly when complementing technological and structural change. In general, higher consumption lifestyles have a greater environmental impact. Overall, food accounts for the largest share of consumption-based GHG emissions with nearly 20% of the global carbon footprint, followed by housing, mobility, services, manufactured products, and construction. Food and services are more significant in poor countries, while mobility and manufactured goods are more significant in rich countries.

Dietary Change

A 2014 study into the real-life diets of British people estimates their greenhouse gas contributions (CO_2eq) to be: 7.19 kg/day for high meat-eaters through to 3.81 kg/day for vegetarians and 2.89 kg/day for vegans. The widespread adoption of a vegetarian diet could cut food-related greenhouse gas emissions by 63% by 2050. China introduced new dietary guidelines in 2016 which aim to cut meat consumption by 50% and thereby reduce greenhouse gas emissions by 1 billion tonnes by 2030. A 2016 study concluded that taxes on meat and milk could simultaneously result in reduced greenhouse gas emissions and healthier diets. The study analyzed surcharges of 40% on beef and 20% on milk and suggests that an optimum plan would reduce emissions by 1 billion tonnes per year.

Sinks and Negative Emissions

A carbon sink is a natural or artificial reservoir that accumulates and stores some carbon-containing chemical compound for an indefinite period, such as a growing forest. A negative carbon dioxide emission on the other hand is a permanent removal of carbon dioxide out of the atmosphere, such as directly capturing carbon dioxide in the atmosphere and storing it in geologic formations underground.

The Antarctic Climate and Ecosystems Cooperative Research Centre (ACE-CRC) notes that one third of humankind's annual emissions of CO_2 are absorbed by the oceans. However, this also leads to ocean acidification, with potentially significant impacts on marine life. Acidification lowers the level of carbonate ions available for calcifying organisms to form their shells. These organisms include plankton species that contribute to the foundation of the Southern Ocean food web. However acidification may impact on a broad range of other physiological and ecological processes, such as fish respiration, larval development and changes in the solubility of both nutrients and toxins.

Reforestation and Afforestation

Transferring land rights to indigenous inhabitants is argued to efficiently conserve forests. Regrowth of forests on abandoned farmland restores more forest than that lost to deforestation.

Almost 20 percent (8 $GtCO_2$/year) of total greenhouse-gas emissions were from deforestation in 2007. It is estimated that avoided deforestation reduces CO_2 emissions at a rate of 1 tonne of CO_2 per $1–5 in opportunity costs from lost agriculture. Reforestation and afforestation, where there was previously no forest, could save at least another 1 $GtCO_2$/year, at an estimated cost of $5–15/$tCO_2$.

Transferring rights over land from public domain to its indigenous inhabitants is argued to be a cost effective strategy to conserve forests. This includes the protection of such rights entitled in existing laws, such as India's Forest Rights Act. The transferring of such rights in China, perhaps the largest land reform in modern times, has been argued to have increased forest cover. In Brazil, forested areas given tenure to indigenous groups have even lower rates of clearing than national parks. A 2016 report concludes that modest investments in indigenous land rights will generate economic, social, and environmental returns for the communities involved and for climate protection. The report quantifies the economic value of securing such rights, with a focus on the Amazon region.

With increased intensive agriculture and urbanization, there is an increase in the amount of abandoned farmland. By some estimates, for every half a hectare of original old-growth forest cut down, more than 20 hectares of new secondary forests are growing, even though they do not have the same biodiversity as the original forests and original forests store 60% more carbon than these new secondary forests. According to a study in *Science*, promoting regrowth on abandoned farmland could offset years of carbon emissions.

It appears that attempts to reduce atmospheric carbon through biomass absorption coupled with sequestration can succeed only if combined with early and far-reaching reduction of current greenhouse gas generation by all means available. Absent such reduction, pursuit of biomass absorption plus sequestration as a late-regret measure would require prohibitively massive use of resources in order to succeed.

Avoided Desertification

Restoring grasslands store CO_2 from the air into plant material. Grazing livestock, usually not left to wander, would eat the grass and would minimize any grass growth. However, grass left alone would eventually grow to cover its own growing buds, preventing them from photosynthesizing and the dying plant would stay in place. A method proposed to restore grasslands uses fences with many small paddocks and moving herds from one paddock to another after a day a two in order to mimick natural grazers and allowing the grass to grow optimally. Additionally, when part of leaf matter is consumed by a herding animal, a corresponding amount of root matter is sloughed off too as it would not be able to sustain the previous amount of root matter and while most of the lost root matter would rot and enter the atmosphere, part of the carbon is sequestered into the soil. It is estimated that increasing the carbon content of the soils in the world's 3.5 billion hectares of agricultural grassland by 1% would offset nearly 12 years of CO_2 emissions. Allan Savory, as part of holistic management, claims that while large herds are often blamed for desertification, prehistoric lands supported large or larger herds and areas where herds were removed in the United States are still desertifying.

Managed grazing methods are argued to be able to restore grasslands, thereby significantly decreasing atmospheric CO_2 levels.

Carbon Capture and Storage

Carbon capture and storage (CCS) is a method to mitigate climate change by capturing carbon dioxide (CO_2) from large point sources such as power plants and subsequently storing it away safely instead of releasing it into the atmosphere. The Intergovernmental Panel on Climate Change says CCS could contribute between 10% and 55% of the cumulative worldwide carbon-mitigation effort over the next 90 years. The International Energy Agency says CCS is "the most important single new technology for CO_2 savings" in power generation and industry. Though it requires up to 40% more energy to run a CCS coal power plant than a regular coal plant, CCS could potentially capture about 90% of all the carbon emitted by the plant. Norway, which first began storing CO_2, has cut its emissions by almost a million tons a year, or about 3% of the country's 1990 levels. As of late 2011, the total CO_2 storage capacity of all 14 projects in operation or under construction is over 33 million tonnes a year. This is broadly equivalent to preventing the emissions from more than six million cars from entering the atmosphere each year.

Negative Carbon Dioxide Emissions

Creating negative carbon dioxide emissions literally removes carbon from the atmosphere. Examples are direct air capture, biochar, bio-energy with carbon capture and storage and enhanced weathering technologies. These processes are sometimes considered as variations of sinks or mitigation, and sometimes as geoengineering.

In combination with other mitigation measures, sinks in combination with negative carbon emissions are considered crucial for meeting the 350 ppm target, and even the less conservative 450 ppm target.

Geoengineering

Geoengineering is seen by some as an alternative to mitigation and adaptation, but by others as an entirely separate response to climate change. In a literature assessment, Barker *et al.* (2007) described geoengineering as a type of mitigation policy. IPCC (2007) concluded that geoengineering options, such as ocean fertilization to remove CO_2 from the atmosphere, remained largely unproven. It was judged that reliable cost estimates for geoengineering had not yet been published.

Chapter 28 of the National Academy of Sciences report *Policy Implications of Greenhouse Warming: Mitigation, Adaptation, and the Science Base* (1992) defined geoengineering as "options that would involve large-scale engineering of our environment in order to combat or counteract the effects of changes in atmospheric chemistry." They evaluated a range of options to try to give preliminary answers to two questions: can these options work and could they be carried out with a reasonable cost. They also sought to encourage discussion of a third question — what adverse side effects might there be. The following types of option were examined: reforestation, increasing ocean absorption of carbon dioxide (carbon sequestration) and screening out some sunlight. NAS also argued "Engineered countermeasures need to be evaluated but should not be implemented without broad understanding of the direct effects and the potential side effects, the ethical issues, and the risks.". In July 2011 a report by the United States Government Accountability Office on geoengineering found that "[c]limate engineering technologies do not now offer a viable response to global climate change."

Carbon Dioxide Removal

Carbon dioxide removal has been proposed as a method of reducing the amount of radiative forcing. A variety of means of artificially capturing and storing carbon, as well as of enhancing natural sequestration processes, are being explored. The main natural process is photosynthesis by plants and single-celled organisms. Artificial processes vary, and concerns have been expressed about the long-term effects of some of these processes.

It is notable that the availability of cheap energy and appropriate sites for geological storage of carbon may make carbon dioxide air capture viable commercially. It is, however, generally expected that carbon dioxide air capture may be uneconomic when compared to carbon capture and storage from major sources — in particular, fossil fuel powered power stations, refineries, etc. In such cases, costs of energy produced will grow significantly. However, captured CO_2 can be used to force more crude oil out of oil fields, as Statoil and Shell have made plans to do. CO_2 can also be used in commercial greenhouses, giving an opportunity to kick-start the technology. Some attempts have been made to use algae to capture smokestack emissions, notably the GreenFuel Technologies Corporation, who have now shut down operations.

Solar Radiation Management

The main purpose of solar radiation management seek to reflect sunlight and thus reduce global warming. The ability of stratospheric sulfate aerosols to create a global dimming effect has made them a possible candidate for use in climate engineering projects.

Non-CO_2 Greenhouse Gases

CO_2 is not the only GHG relevant to mitigation, and governments have acted to regulate the emissions of other GHGs emitted by human activities (anthropogenic GHGs). The emissions caps agreed to by most developed countries under the Kyoto Protocol regulate the emissions of almost all the anthropogenic GHGs. These gases are CO_2, methane (CH_4), nitrous oxide (N_2O), the hydrofluorocarbons (HFC), perfluorocarbons (PFC), and sulfur hexafluoride (SF_6).

Stabilizing the atmospheric concentrations of the different anthropogenic GHGs requires an understanding of their different physical properties. Stabilization depends both on how quickly GHGs are added to the atmosphere and how fast they are removed. The rate of removal is measured by the atmospheric lifetime of the GHG in question. Here, the lifetime is defined as the time required for a given perturbation of the GHG in the atmosphere to be reduced to 37% of its initial amount. Methane has a relatively short atmospheric lifetime of about 12 years, while N_2O's lifetime is about 110 years. For methane, a reduction of about 30% below current emission levels would lead to a stabilization in its atmospheric concentration, while for N_2O, an emissions reduction of more than 50% would be required.

Methane is a significantly more potent greenhouse gas than carbon dioxide in the amount of heat it can trap, especially in the short term. Burning one molecule of methane generates one molecule of carbon dioxide, indicating there may be no net benefit in using gas as a fuel source. Reducing the amount of waste methane produced in the first place and moving away from use of gas as a fuel source will have a greater beneficial impact, as might other approaches to productive use of otherwise-wasted methane. In terms of prevention, vaccines are being developed in Australia to reduce the significant global warming contributions from methane released by livestock via flatulence and eructation.

Another physical property of the anthropogenic GHGs relevant to mitigation is the different abilities of the gases to trap heat (in the form of infrared radiation). Some gases are more effective at trapping heat than others, e.g., SF_6 is 22,200 times more effective a GHG than CO_2 on a per-kilogram basis. A measure for this physical property is the global warming potential (GWP), and is used in the Kyoto Protocol.

Although not designed for this purpose, the Montreal Protocol has probably benefited climate change mitigation efforts. The Montreal Protocol is an international treaty that has successfully reduced emissions of ozone-depleting substances (for example, CFCs), which are also greenhouse gases.

By Sector

Transport

Transportation emissions account for roughly 1/4 of emissions worldwide, and are even more important in terms of impact in developed nations especially in North America and Australia. Many citizens of countries like the United States and Canada who drive personal cars often, see well over half of their climate change impact stemming from the emissions produced from their cars. Modes of mass transportation such as bus, light rail (metro, subway, etc.), and long-distance rail are far and away the most energy-efficient means of motorized transportation for passengers, able to use in many cases over twenty times less energy per person-distance than a personal automobile. Modern energy-efficient technologies, such as plug-in hybrid electric vehicles and carbon-neutral synthetic gasoline & Jet fuel may also help to reduce the consumption of petroleum, land use changes and emissions of carbon dioxide. Utilizing rail transport, especially electric rail, over the far less efficient air transport and truck transport significantly reduces emissions. With the use of electric trains and cars in transportation there is the opportunity to run them with low-carbon power, producing far fewer emissions.

The Tesla Roadster emits no tailpipe emissions, uses lithium ion batteries to achieve 220 mi (350 km) per charge, while also capable of going 0–60 in under 4 seconds.

Bicycles have almost no carbon footprint compared to cars, and canal transport may represent a positive option for certain types of freight in the 21st century.

Urban Planning

Effective urban planning to reduce sprawl aims to decrease Vehicle Miles Travelled (VMT), lowering emissions from transportation. Personal cars are extremely inefficient at moving passengers, while public transport and bicycles are many times more efficient (as is the simplest form of human transportation, walking). All of these are encouraged by urban/community planning and are an effective way to reduce greenhouse gas emissions. Between 1982 and 1997, the amount of land consumed for urban development in the United States increased by 47 percent while the nation's population grew by only 17 percent. Inefficient land use development practices have increased infrastructure costs as well as the amount of energy needed for transportation, community services, and buildings.

At the same time, a growing number of citizens and government officials have begun advocating a smarter approach to land use planning. These smart growth practices include compact community development, multiple transportation choices, mixed land uses, and practices to conserve green space. These programs offer environmental, economic, and quality-of-life benefits; and they also serve to reduce energy usage and greenhouse gas emissions.

Approaches such as New Urbanism and Transit-oriented development seek to reduce distances travelled, especially by private vehicles, encourage public transit and make walking and cycling more attractive options. This is achieved through "medium-density", mixed-use planning and the concentration of housing within walking distance of town centers and transport nodes.

Smarter growth land use policies have both a direct and indirect effect on energy consuming behavior. For example, transportation energy usage, the number one user of petroleum fuels, could be significantly reduced through more compact and mixed use land development patterns, which in turn could be served by a greater variety of non-automotive based transportation choices.

Building Design

Emissions from housing are substantial, and government-supported energy efficiency programmes can make a difference.

For institutions of higher learning in the United States, greenhouse gas emissions depend primarily on total area of buildings and secondarily on climate. If climate is not taken into account, annual greenhouse gas emissions due to energy consumed on campuses plus purchased electricity can be estimated with the formula, $E=aS^b$, where a =0.001621 metric tonnes of CO_2 equivalent/square foot or 0.0241 metric tonnes of CO_2 equivalent/square meter and b = 1.1354.

New buildings can be constructed using passive solar building design, low-energy building, or zero-energy building techniques, using renewable heat sources. Existing buildings can be made more efficient through the use of insulation, high-efficiency appliances (particularly hot water heaters and furnaces), double- or triple-glazed gas-filled windows, external window shades, and building orientation and siting. Renewable heat sources such as shallow geothermal and passive solar energy reduce the amount of greenhouse gasses emitted. In addition to designing buildings which are more energy-efficient to heat, it is possible to design buildings that are more energy-efficient to cool by using lighter-coloured, more reflective materials in the development of urban areas (e.g. by painting roofs white) and planting trees. This saves energy because it cools buildings and reduces the urban heat island effect thus reducing the use of air conditioning.

Agriculture

According to the EPA, agricultural soil management practices can lead to production and emission of nitrous oxide (N_2O), a major greenhouse gas and air pollutant. Activities that can contribute to N_2O emissions include fertilizer usage, irrigation and tillage. The management of soils accounts for over half of the emissions from the Agriculture sector. Cattle livestocks account for one third of emissions, through methane emissions. Manure management and rice cultivation also produce gaseous emissions.

Methods that significantly enhance carbon sequestration in soil include no-till farming, residue mulching, cover cropping, and crop rotation, all of which are more widely used in organic farming than in conventional farming. Because only 5% of US farmland currently uses no-till and residue mulching, there is a large potential for carbon sequestration.

A 2015 study found that farming can deplete soil carbon and render soil incapable of supporting life; however, the study also showed that conservation farming can protect carbon in soils, and repair damage over time.

The farming practise of cover crops has been recognized as climate-smart agriculture by the White House.

Societal Controls

Another method being examined is to make carbon a new currency by introducing tradeable "personal carbon credits". The idea being it will encourage and motivate individuals to reduce their 'carbon footprint' by the way they live. Each citizen will receive a free annual quota of carbon that they can use to travel, buy food, and go about their business. It has been suggested that by using this concept it could actually solve two problems; pollution and poverty, old age pensioners will actually be better off because they fly less often, so they can cash in their quota at the end of the year to pay heating bills and so forth.

Population

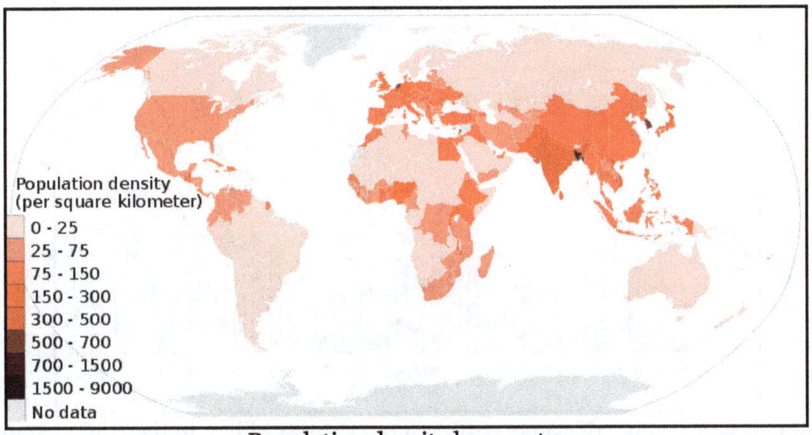

Population density by country.

Various organizations promote population control as a means for mitigating global warming. Proposed measures include improving access to family planning and reproductive health care and information, reducing natalistic politics, public education about the consequences of continued population growth, and improving access of women to education and economic opportunities.

Population control efforts are impeded by there being somewhat of a taboo in some countries against considering any such efforts. Also, various religions discourage or prohibit some or all forms of birth control.

Population size has a different per capita effect on global warming in different countries, since the per capita production of anthropogenic greenhouse gases varies greatly by country.

Costs and Benefits

Costs

The Stern Review proposes stabilising the concentration of greenhouse-gas emissions in the atmosphere at a maximum of 550ppm CO_2e by 2050. The Review estimates that this would mean cutting total greenhouse-gas emissions to three quarters of 2007 levels. The Review further estimates that the cost of these cuts would be in the range −1.0 to +3.5% of World GDP, (i.e. GWP), with an average estimate of approximately 1%. Stern has since revised his estimate to 2% of GWP. For comparison, the Gross World Product (GWP) at PPP was estimated at $74.5 trillion in 2010, thus 2% is approximately $1.5 trillion. The Review emphasises that these costs are contingent on steady reductions in the cost of low-carbon technologies. Mitigation costs will also vary according to how and when emissions are cut: early, well-planned action will minimise the costs.

One way of estimating the cost of reducing emissions is by considering the likely costs of potential technological and output changes. Policy makers can compare the marginal abatement costs of different methods to assess the cost and amount of possible abatement over time. The marginal abatement costs of the various measures will differ by country, by sector, and over time.

Benefits

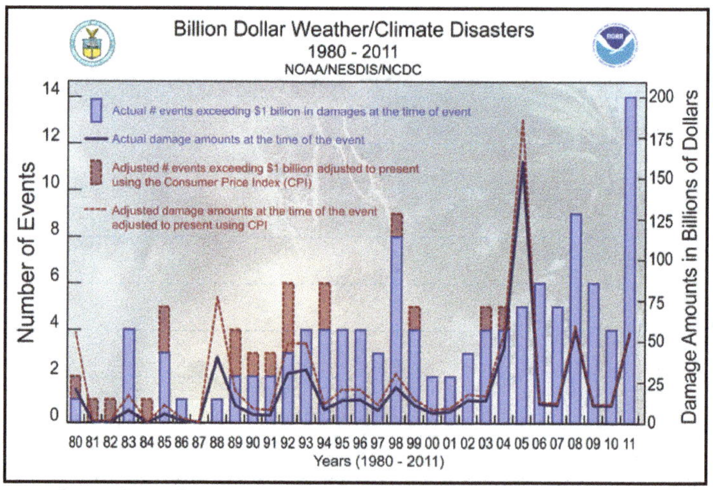

Total extreme weather cost and number of events costing more than $1 billion in the United States from 1980 to 2011.

Yohe *et al.* (2007) assessed the literature on sustainability and climate change. With high confidence, they suggested that up to the year 2050, an effort to cap greenhouse gas (GHG) emissions at 550 ppm would benefit developing countries significantly. This was judged to be especially the case when combined with enhanced adaptation. By 2100, however, it was still judged likely that there would be significant effects of global warming. This was judged to be the case even with aggressive mitigation and significantly enhanced adaptive capacity.

Sharing

One of the aspects of mitigation is how to share the costs and benefits of mitigation policies. There is no scientific consensus over how to share these costs and benefits (Toth *et al.*, 2001). In terms of the politics of mitigation, the UNFCCC's ultimate objective is to stabilize concentrations of GHG in the atmosphere at a level that would prevent "dangerous" climate change (Rogner *et al.*, 2007).

GHG emissions are an important correlate of wealth, at least at present (Banuri *et al.*, 1996, pp. 91–92). Wealth, as measured by per capita income (i.e., income per head of population), varies widely between different countries. Activities of the poor that involve emissions of GHGs are often associated with basic needs, such as heating to stay tolerably warm. In richer countries, emissions tend to be associated with things like cars, central heating, etc. The impacts of cutting emissions could therefore have different impacts on human welfare according to wealth.

Distributing Emissions Abatement Costs

There have been different proposals on how to allocate responsibility for cutting emissions (Banuri *et al.*, 1996, pp. 103–105):

- Egalitarianism: this system interprets the problem as one where each person has equal rights to a global resource, i.e., polluting the atmosphere.
- Basic needs: this system would have emissions allocated according to basic needs, as defined according to a minimum level of consumption. Consumption above basic needs would require countries to buy more emission rights. From this viewpoint, developing countries would need to be at least as well off under an emissions control regime as they would be outside the regime.
- Proportionality and polluter-pays principle: Proportionality reflects the ancient Aristotelian principle that people should receive in proportion to what they put in, and pay in proportion to the damages they cause. This has a potential relationship with the "polluter-pays principle", which can be interpreted in a number of ways:
    - *Historical responsibilities*: this asserts that allocation of emission rights should be based on patterns of past emissions. Two-thirds of the stock of GHGs in the atmosphere at present is due to the past actions of developed countries (Goldemberg *et al.*, 1996, p. 29).
    - *Comparable burdens and ability to pay*: with this approach, countries would reduce emissions based on comparable burdens and their ability to take on the costs of reduction. Ways to assess burdens include monetary costs per head of population, as well as other, more complex measures, like the UNDP's Human Development Index.
    - *Willingness to pay*: with this approach, countries take on emission reductions based on their ability to pay along with how much they benefit from reducing their emissions.

Specific Proposals

- Ad hoc: Lashof (1992) and Cline (1992), for example, suggested that allocations based partly on GNP could be a way of sharing the burdens of emission reductions. This is because GNP and economic activity are partially tied to carbon emissions.
- Equal per capita entitlements: this is the most widely cited method of distributing abatement costs, and is derived from egalitarianism (Banuri *et al.*, 1996, pp. 106–107). This approach can be divided into two categories. In the first category, emissions are allocated

according to national population. In the second category, emissions are allocated in a way that attempts to account for historical (cumulative) emissions.

- Status quo: with this approach, historical emissions are ignored, and current emission levels are taken as a status quo right to emit. An analogy for this approach can be made with fisheries, which is a common, limited resource. The analogy would be with the atmosphere, which can be viewed as an exhaustible natural resource. In international law, one state recognized the long-established use of another state's use of the fisheries resource. It was also recognized by the state that part of the other state's economy was dependent on that resource.

Governmental and Intergovernmental Action

Many countries, both developing and developed, are aiming to use cleaner technologies. Use of these technologies aids mitigation and could result in substantial reductions in CO_2 emissions. Policies include targets for emissions reductions, increased use of renewable energy, and increased energy efficiency. It is often argued that the results of climate change are more damaging in poor nations, where infrastructures are weak and few social services exist. The Commitment to Development Index is one attempt to analyze rich country policies taken to reduce their disproportionate use of the global commons. Countries do well if their greenhouse gas emissions are falling, if their gas taxes are high, if they do not subsidize the fishing industry, if they have a low fossil fuel rate per capita, and if they control imports of illegally cut tropical timber.

Kyoto Protocol

The main current international agreement on combating climate change is the Kyoto Protocol, which came into force on 16 February 2005. The Kyoto Protocol is an amendment to the United Nations Framework Convention on Climate Change (UNFCCC). Countries that have ratified this protocol have committed to reduce their emissions of carbon dioxide and five other greenhouse gases, or engage in emissions trading if they maintain or increase emissions of these gases.

Temperature Targets

Actions to mitigate climate change are sometimes based on the goal of achieving a particular temperature target. One of the targets that has been suggested is to limit the future increase in global mean temperature (global warming) to below 2 °C, relative to the pre-industrial level. The 2 °C target was adopted in 2010 by Parties to the United Nations Framework Convention on Climate Change. Most countries of the world are Parties to the UNFCCC. The target had been adopted in 1996 by the European Union Council.

Feasibility of 2 °C

Temperatures have increased by 0.8 °C compared to the pre-industrial level, and another 0.5–0.7 °C is already committed. The 2 °C rise is typically associated in climate models with a carbon dioxide equivalent concentration of 400–500 ppm by volume; the current (January 2015) level of carbon dioxide alone is 400 ppm by volume, and rising at 1–3 ppm annually. Hence, to avoid a very likely breach of the 2 °C target, CO_2 levels would have to be stabilised very soon; this is generally regarded as unlikely, based on current programs in place to date. The importance of change is illustrated by the fact that world economic energy efficiency is improving at only half the rate of world economic growth.

Views in the Literature

There is disagreement among experts over whether or not the 2 °C target can be met. For example, according to Anderson and Bows (2011), "there is little to no chance" of meeting the target. On the other hand, according to Alcamo *et al.* (2013):

- Policies adopted by parties to the UNFCCC are too weak to meet a 2 or 1.5 °C target. However, these targets might still be achievable if more stringent mitigation policies are adopted immediately.
- Cost-effective 2 °C scenarios project annual global greenhouse gas emissions to peak before the year 2020, with deep cuts in emissions thereafter, leading to a reduction in 2050 of 41% compared to 1990 levels.

Discussion on other Targets

Scientific analysis can provide information on the impacts of climate change and associated policies, such as reducing GHG emissions. However, deciding what policies are best requires value judgements. For example, limiting global warming to 1 °C relative to pre-industrial levels may help to reduce climate change damages more than a 2 °C limit. However, a 1 °C limit may be more costly to achieve than a 2 °C limit.

According to some analysts, the 2 °C "guardrail" is inadequate for the needed degree and timeliness of mitigation. On the other hand, some economic studies suggest more modest mitigation policies. For example, the emissions reductions proposed by Nordhaus (2010) might lead to global warming (in the year 2100) of around 3 °C, relative to pre-industrial levels.

Official Long-term Target of 1.5 °C

In 2015, two official UNFCCC scientific expert bodies came to the conclusion that, "in some regions and vulnerable ecosystems, high risks are projected even for warming above 1.5°C". This expert position was, together with the strong diplomatic voice of the poorest countries and the island nations in the Pacific, the driving force leading to the decision of the Paris Conference 2015, to lay down this 1.5 °C long-term target on top of the existing 2 °C goal.

Encouraging Use Changes

Emissions Tax

An emissions tax on greenhouse gas emissions requires individual emitters to pay a fee, charge or tax for every tonne of greenhouse gas released into the atmosphere. Most environmentally related taxes with implications for greenhouse gas emissions in OECD countries are levied on energy products and motor vehicles, rather than on CO_2 emissions directly.

Emission taxes can be both cost-effective and environmentally effective. Difficulties with emission taxes include their potential unpopularity, and the fact that they cannot guarantee a particular level of emissions reduction. Emissions or energy taxes also often fall disproportionately on lower income classes. In developing countries, institutions may be insufficiently developed for the collection of emissions fees from a wide variety of sources.

Subsidies

According to Mark Z. Jacobson, a program of subsidization balanced against expected flood costs could pay for conversion to 100% renewable power by 2030. Jacobson, and his colleague Mark Delucchi, suggest that the cost to generate and transmit power in 2020 will be less than 4 cents per kilowatt hour (in 2007 dollars) for wind, about 4 cents for wave and hydroelectric, from 4 to 7 cents for geothermal, and 8 cents per kWh for solar, fossil, and nuclear power.

Investment

Another indirect method of encouraging uses of renewable energy, and pursue sustainability and environmental protection, is that of prompting investment in this area through legal means, something that is already being done at national level as well as in the field of international investment.

Carbon Emissions Trading

With the creation of a market for trading carbon dioxide emissions within the Kyoto Protocol, it is likely that London financial markets will be the centre for this potentially highly lucrative business; the New York and Chicago stock markets may have a lower trade volume than expected as long as the US maintains its rejection of the Kyoto.

However, emissions trading may delay the phase-out of fossil fuels.

In the north-east United States, a successful cap and trade program has shown potential for this solution.

The European Union Emission Trading Scheme (EU ETS) is the largest multi-national, greenhouse gas emissions trading scheme in the world. It commenced operation on 1 January 2005, and all 28 member states of the European Union participate in the scheme which has created a new market in carbon dioxide allowances estimated at 35 billion Euros (US$43 billion) per year. The Chicago Climate Exchange was the first (voluntary) emissions market, and is soon to be followed by Asia's first market (Asia Carbon Exchange). A total of 107 million metric tonnes of carbon dioxide equivalent have been exchanged through projects in 2004, a 38% increase relative to 2003 (78 Mt CO_2e).

Twenty three multinational corporations have come together in the G8 Climate Change Roundtable, a business group formed at the January 2005 World Economic Forum. The group includes Ford, Toyota, British Airways and BP. On 9 June 2005 the Group published a statement stating that there was a need to act on climate change and claiming that market-based solutions can help. It called on governments to establish "clear, transparent, and consistent price signals" through "creation of a long-term policy framework" that would include all major producers of greenhouse gases.

The Regional Greenhouse Gas Initiative is a proposed carbon trading scheme being created by nine North-eastern and Mid-Atlantic American states; Connecticut, Delaware, Maine, Massachusetts, New Hampshire, New Jersey, New York, Rhode Island and Vermont. The scheme was due to be developed by April 2005 but has not yet been completed.

Implementation

Implementation puts into effect climate change mitigation strategies and targets. These can be targets set by international bodies or voluntary action by individuals or institutions. This is the most important, expensive and least appealing aspect of environmental governance.

Funding

Implementation requires funding sources but is often beset by disputes over who should provide funds and under what conditions. A lack of funding can be a barrier to successful strategies as there are no formal arrangements to finance climate change development and implementation. Funding is often provided by nations, groups of nations and increasingly NGO and private sources. These funds are often channelled through the Global Environmental Facility (GEF). This is an environmental funding mechanism in the World Bank which is designed to deal with global environmental issues. The GEF was originally designed to tackle four main areas: biological diversity, climate change, international waters and ozone layer depletion, to which land degradation and persistent organic pollutant were added. The GEF funds projects that are agreed to achieve global environmental benefits that are endorsed by governments and screened by one of the GEF's implementing agencies.

Problems

There are numerous issues which result in a current perceived lack of implementation. It has been suggested that the main barriers to implementation are, Uncertainty, Fragmentation, Institutional void, Short time horizon of policies and politicians and Missing motives and willingness to start adapting. The relationships between many climatic processes can cause large levels of uncertainty as they are not fully understood and can be a barrier to implementation. When information on climate change is held between the large numbers of actors involved it can be highly dispersed, context specific or difficult to access causing fragmentation to be a barrier. Institutional void is the lack of commonly accepted rules and norms for policy processes to take place, calling into question the legitimacy and efficacy of policy processes. The Short time horizon of policies and politicians often means that climate change policies are not implemented in favour of socially favoured societal issues. Statements are often posed to keep the illusion of political action to prevent or postpone decisions being made. Missing motives and willingness to start adapting is a large barrier as it prevents any implementation.

The issues that arise with a system which involves international government cooperation, such as Cap and Trade, could potentially be improved with a polycentric approach where the rules are enforced by many small sections of authority as apposed to one overall enforcement agency.

Occurrence

Despite a perceived lack of occurrence, evidence of implementation is emerging internationally. Some examples of this are the initiation of NAPA's and of joint implementation. Many developing nations have made National Adaptation Programs of Action (NAPAs) which are frameworks to prioritize adaption needs. The implementation of many of these is supported by GEF agencies. Many developed countries are implementing 'first generation' institutional adaption plans particularly at the state and local government scale. There has also been a push towards joint implementation between countries by the UNFCC as this has been suggested as a cost-effective way for objectives to be achieved.

United States

Efforts to reduce greenhouse gas emissions by the United States include energy policies which encourage efficiency through programs like Energy Star, Commercial Building Integration, and the Industrial Technologies Program. On 12 November 1998, Vice President Al Gore symbolically signed the Kyoto Protocol, but he indicated participation by the developing nations was necessary prior its being submitted for ratification by the United States Senate.

In 2007, Transportation Secretary Mary Peters, with White House approval, urged governors and dozens of members of the House of Representatives to block California's first-in-the-nation limits on greenhouse gases from cars and trucks, according to e-mails obtained by Congress. The US Climate Change Science Program is a group of about twenty federal agencies and US Cabinet Departments, all working together to address global warming.

The Bush administration pressured American scientists to suppress discussion of global warming, according to the testimony of the Union of Concerned Scientists to the Oversight and Government Reform Committee of the US House of Representatives. "High-quality science" was "struggling to get out," as the Bush administration pressured scientists to tailor their writings on global warming to fit the Bush administration's skepticism, in some cases at the behest of an ex-oil industry lobbyist. "Nearly half of all respondents perceived or personally experienced pressure to eliminate the words 'climate change,' 'global warming' or other similar terms from a variety of communications." Similarly, according to the testimony of senior officers of the Government Accountability Project, the White House attempted to bury the report "National Assessment of the Potential Consequences of Climate Variability and Change," produced by US scientists pursuant to US law. Some US scientists resigned their jobs rather than give in to White House pressure to underreport global warming.

In the absence of substantial federal action, state governments have adopted emissions-control laws such as the Regional Greenhouse Gas Initiative in the Northeast and the Global Warming Solutions Act of 2006 in California.

Developing Countries

In order to reconcile economic development with mitigating carbon emissions, developing countries need particular support, both financial and technical. One of the means of achieving this is the Kyoto Protocol's Clean Development Mechanism (CDM). The World Bank's Prototype Carbon Fund is a public private partnership that operates within the CDM.

An important point of contention, however, is how overseas development assistance not directly related to climate change mitigation is affected by funds provided to climate change mitigation. One of the outcomes of the UNFCC Copenhagen Climate Conference was the Copenhagen Accord, in which developed countries promised to provide US$30 million between 2010 and 2012 of new and additional resources. Yet it remains unclear what exactly the definition of additional is and the European Commission has requested its member states to define what they understand to be additional, and researchers at the Overseas Development Institute have found four main understandings:

1. Climate finance classified as aid, but additional to (over and above) the '0.7%' ODA target;
2. Increase on previous year's Official Development Assistance (ODA) spent on climate change mitigation;
3. Rising ODA levels that include climate change finance but where it is limited to a specified percentage;
4. Increase in climate finance not connected to ODA.

The main point being that there is a conflict between the OECD states budget deficit cuts, the need to help developing countries adapt to develop sustainably and the need to ensure that funding does not come from cutting aid to other important Millennium Development Goals.

However, none of these initiatives suggest a quantitative cap on the emissions from developing countries. This is considered as a particularly difficult policy proposal as the economic growth of developing countries are proportionally reflected in the growth of greenhouse emissions. Critics of mitigation often argue that, the developing countries' drive to attain a comparable living standard to the developed countries would doom the attempt at mitigation of global warming. Critics also argue that holding down emissions would shift the human cost of global warming from a general one to one that was borne most heavily by the poorest populations on the planet.

In an attempt to provide more opportunities for developing countries to adapt clean technologies, UNEP and WTO urged the international community to reduce trade barriers and to conclude the Doha trade round "which includes opening trade in environmental goods and services".

Non-governmental Approaches

While many of the proposed methods of mitigating global warming require governmental funding, legislation and regulatory action, individuals and businesses can also play a part in the mitigation effort.

Choices in Personal Actions and Business Operations

Environmental groups encourage individual action against global warming, often aimed at the consumer. Common recommendations include lowering home heating and cooling usage, burning less gasoline, supporting renewable energy sources, buying local products to reduce transportation, turning off unused devices, and various others.

A geophysicist at Utrecht University has urged similar institutions to hold the vanguard in voluntary mitigation, suggesting the use of communications technologies such as videoconferencing to reduce their dependence on long-haul flights.

Air Travel and Shipment

In 2008, climate scientist Kevin Anderson raised concern about the growing effect of rapidly increasing global air transport on the climate in a paper, and a presentation, suggesting that reversing this trend is necessary to reduce emissions.

Part of the difficulty is that when aviation emissions are made at high altitude, the climate impacts are much greater than otherwise. Others have been raising the related concerns of the increasing hypermobility of individuals, whether traveling for business or pleasure, involving frequent and often long distance air travel, as well as air shipment of goods.

Business Opportunities and Risks

On 9 May 2005 Jeff Immelt, the chief executive of General Electric (GE), announced plans to reduce GE's global warming related emissions by one percent by 2012. "GE said that given its projected growth, those emissions would have risen by 40 percent without such action."

On 21 June 2005 a group of leading airlines, airports and aerospace manufacturers pledged to work together to reduce the negative environmental impact of aviation, including limiting the impact of air travel on climate change by improving fuel efficiency and reducing carbon dioxide emissions of new aircraft by fifty percent per seat kilometre by 2020 from 2000 levels. The group aims to develop a common reporting system for carbon dioxide emissions per aircraft by the end of 2005, and pressed for the early inclusion of aviation in the European Union's carbon emission trading scheme.

Investor Response

Climate change is also a concern for large institutional investors who have a long term time horizon and potentially large exposure to the negative impacts of global warming because of the large geographic footprint of their multi-national holdings. SRI (Socially responsible investing) Funds allow investors to invest in funds that meet high ESG (environmental, social, governance) standards as such funds invest in companies that are aligned with these goals. Proxy firms can be used to draft guidelines for investment managers that take these concerns into account.

Legal Action

In some countries, those affected by climate change may be able to sue major producers. Attempts at litigation have been initiated by entire peoples such as Palau and the Inuit, as well as non-governmental organizations such as the Sierra Club. Although proving that particular weather events are due specifically to global warming may never be possible, methodologies have been developed to show the increased risk of such events caused by global warming.

For a legal action for negligence (or similar) to succeed, "Plaintiffs must show that, more probably than not, their individual injuries were caused by the risk factor in question, as opposed to any other cause. This has sometimes been translated to a requirement of a relative risk of at least two." Another route (though with little legal bite) is the World Heritage Convention, if it can be shown that climate change is affecting World Heritage Sites like Mount Everest.

Besides countries suing one another, there are also cases where people in a country have taken legal steps against their own government. Legal action for instance has been taken to try to force the US Environmental Protection Agency to regulate greenhouse gas emissions under the Clean Air Act, and against the Export-Import Bank and OPIC for failing to assess environmental impacts (including global warming impacts) under NEPA.

In the Netherlands and Belgium, organisations as Urgenda and the vzw Klimaatzaak in Belgium have also sued their governments as they believe their governments aren't meeting the emission reductions they agreed to. Urgenda has all ready won their case against the Dutch government.

According to a 2004 study commissioned by Friends of the Earth, ExxonMobil and its predecessors caused 4.7 to 5.3 percent of the world's man-made carbon dioxide emissions between 1882 and 2002. The group suggested that such studies could form the basis for eventual legal action.

In 2015, Exxon, received a subpoena. According to the Washington Post and confirmed by the company, the attorney general of New York, Eric Schneiderman, opened an investigation into the possibility that the company had mislead the public and investors about the risks of climate change.

References

- Diesendorf, Mark (2009). Climate action: a campaign manual for greenhouse solutions. Sydney: University of New South Wales Press. p. 116. ISBN 978-1-74223-018-4
- Forest Practices Board. 2007. Lodgepole Pine Stand Structure 25 Years after Mountain Pine Beetle Attack. "Archived copy" (PDF). Archived from the original (PDF) on 2008-10-29. Retrieved 2007-05-12
- Buma, B.; Wessman, C. A. (2011). "Disturbance interactions can impact resilience mechanisms of forests". Ecosphere. 2 (5): art64. doi:10.1890/ES11-00038.1
- Taylor A.F, Kuo F.E, Sullivan W.C (2001). Views of Nature and Self Discipline: Evidence from Inner City Children in Journal of Environmental Psychology (2001), vol. 21
- Adam, David (2009-02-18). "Fifth of world carbon emissions soaked up by extra forest growth, scientists find". The Guardian. London. Retrieved 2010-05-22
- Dan Leon."The Jewish National Fund: How the Land Was 'Redeemed': The JNF's historical concept of exclusively Jewish land is wholly anachronistic"; Palestine-Israel Journal, Vol 12 No. 4 & Vol 13 No. 1, 05/06
- Ceballos, G.; Ehrlich, A. H.; Ehrlich, P. R. (2015). The Annihilation of Nature: Human Extinction of Birds and Mammals. Baltimore, Maryland: Johns Hopkins University Press. pp. 146 ISBN 1421417189
- "SAFnet Dictionary | Definition For [afforestation]". Dictionaryofforestry.org. 2008-10-23. Archived from the original on 2012-03-14. Retrieved 2012-02-17
- "Forest Service Chief testifies before Senate appropriations committee on 2013 agency budget". US Forest Service. 18 April 2012. Retrieved 29 April 2012
- Parr JF, Sullivan LA (2005). "Soil carbon sequestration in phytoliths". Soil Biology and Biochemistry. 37: 117–24. doi:10.1016/j.soilbio.2004.06.013
- Stern, Nicholas Herbert (2007). The economics of climate change: the Stern review. Cambridge, UK: Cambridge University Press. p. xxv. ISBN 0-521-70080-9
- MBD Energy Ltd. MBD captures Loy Yang Carbon Emissions. Eco Investor June 2009 "Archived copy". Archived from the original on 2011-07-14. Retrieved 2010-01-27. accessed 28 Jan 2010
- UK Royal Society (September 2009), Geoengineering the climate: science, governance and uncertainty (PDF), London: UK Royal Society, ISBN 978-0-85403-773-5 , RS Policy document 10/09. Report website

- Andrews, Peter (2008). Beyond the brink: Peter Andrews' radical vision for a sustainable Australian landscape. Sydney: ABC Books for the Australian Broadcasting Corporation. p. 40. ISBN 0-7333-2410-X
- K.T. Weber, B.S. Gokhale, (2011). "Effect of grazing on soil-water content in semiarid rangelands of southeast Idaho" Journal of Arid Environments. 75, 464-470
- "Stormwater to Street Trees" (PDF). United States Environmental Protection Agency. United States Environmental Protection Agency. Retrieved September 4, 2015
- Canadell, J.G.; M.R. Raupach (2008-06-13). "Managing Forests for Climate Change". Science. AAAS. 320 (5882): 1456–1457. PMID 18556550. doi:10.1126/science.1155458
- Retallack, Gregory (2001). "Cenozoic Expansion of Grasslands and Climatic Cooling" (PDF). The Journal of Geology. University of Chicago Press. 109: 407–426. Bibcode:2001JG....109..407R. doi:10.1086/320791
- Dale B, Kim S (2004). "Cumulative Energy and Global Warming Impact from the Production of Biomass for Biobased Products". Journal of Industrial Ecology. 7 (3-4): 147–62. doi:10.1162/108819803323059442

Chapter 4

Deforestation

Deforestation is the large scale removal of trees in a forest or woodland area to provide clearings for other human activities such as agricultural croplands, urbanization or mining activities. It has brought about consequences like soil erosion, climate change, biodiversity depletion, etc. Deforestation and its associated features like illegal logging, forest dieback, climate change, etc. are covered in detail in this chapter.

Satellite photograph of deforestation in progress in eastern Bolivia. Worldwide, 10 percent of wilderness areas were lost between 1990 and 2015.

Deforestation, clearance or clearing is the removal of a forest or stand of trees where the land is thereafter converted to a non-forest use. Examples of deforestation include conversion of forestland to farms, ranches, or urban use. Tropical rainforests is where the most concentrated deforestation occurs. About 30% of Earth's land surface is covered by forests.

In temperate mesic climates, natural regeneration of forest stands often will not occur in the absence of disturbance, whether natural or anthropogenic. Furthermore, biodiversity after regeneration harvest often mimics that found after natural disturbance, including biodiversity loss after naturally occurring rainforest destruction.

Deforestation occurs for multiple reasons: trees are cut down to be used or sold as fuel (sometimes in the form of charcoal) or timber, while cleared land is used as pasture for livestock and plantation. The removal of trees without sufficient reforestation has resulted in damage to habitat, biodiversity loss and aridity. It has adverse impacts on biosequestration of atmospheric carbon dioxide. Deforestation has also been used in war to deprive the enemy of cover for its forces and also vital resources. Modern examples of this were the use of Agent Orange by the British military in Malaya during the Malayan Emergency and the United States military in Vietnam during the Vietnam War. As of 2005, net deforestation rates have ceased to increase in countries with a per capita GDP of at least US$4,600. Deforested regions typically incur significant adverse soil erosion and frequently degrade into wasteland.

Disregard of ascribed value, lax forest management and deficient environmental laws are some of the factors that allow deforestation to occur on a large scale. In many countries, deforestation, both naturally occurring and human induced, is an ongoing issue. Deforestation causes extinction, changes to climatic conditions, desertification, and displacement of populations as observed by current conditions and in the past through the fossil record. More than half of all plant and land animal species in the world live in tropical forests.

Between 2000 and 2012, 2.3 million square kilometres (890,000 square miles) of forests around the earth were cut down. As a result of deforestation, only 6.2 million square kilometres (2.4 million square miles) remain of the original 16 million square kilometres (6 million square miles) of forest that formerly covered the earth.

Causes

According to the United Nations Framework Convention on Climate Change (UNFCCC) secretariat, the overwhelming direct cause of deforestation is agriculture. Subsistence farming is responsible for 48% of deforestation; commercial agriculture is responsible for 32% of deforestation; logging is responsible for 14% of deforestation and fuel wood removals make up 5% of deforestation.

Experts do not agree on whether industrial logging is an important contributor to global deforestation. Some argue that poor people are more likely to clear forest because they have no alternatives, others that the poor lack the ability to pay for the materials and labour needed to clear forest. One study found that population increases due to high fertility rates were a primary driver of tropical deforestation in only 8% of cases.

Other causes of contemporary deforestation may include corruption of government institutions, the inequitable distribution of wealth and power, population growth and overpopulation, and urbanization. Globalization is often viewed as another root cause of deforestation, though there are cases in which the impacts of globalization (new flows of labor, capital, commodities, and ideas) have promoted localized forest recovery.

In 2000 the United Nations Food and Agriculture Organization (FAO) found that "the role of population dynamics in a local setting may vary from decisive to negligible," and that deforestation can result from "a combination of population pressure and stagnating economic, social and technological conditions."

The last batch of sawnwood from the peat forest in Indragiri Hulu, Sumatra, Indonesia. Deforestation for oil palm plantation.

The degradation of forest ecosystems has also been traced to economic incentives that make forest conversion appear more profitable than forest conservation. Many important forest functions have no markets, and hence, no economic value that is readily apparent to the forests' owners or the communities that rely on forests for their well-being. From the perspective of the developing world, the benefits of forest as carbon sinks or biodiversity reserves go primarily to richer developed nations and there is insufficient compensation for these services. Developing countries feel that some countries in the developed world, such as the United States of America, cut down their forests centuries ago and benefited greatly from this deforestation, and that it is hypocritical to deny developing countries the same opportunities: that the poor shouldn't have to bear the cost of preservation when the rich created the problem.

Some commentators have noted a shift in the drivers of deforestation over the past 30 years. Whereas deforestation was primarily driven by subsistence activities and government-sponsored development projects like transmigration in countries like Indonesia and colonization in Latin America, India, Java, and so on, during late 19th century and the earlier half of the 20th century. By the 1990s the majority of deforestation was caused by industrial factors, including extractive industries, large-scale cattle ranching, and extensive agriculture.

Problems With Deforestation

Atmospheric

Illegal slash and burn practice in Madagascar, 2010.

Deforestation is ongoing and is shaping climate and geography.

Deforestation is a contributor to global warming, and is often cited as one of the major causes of the enhanced greenhouse effect. Tropical deforestation is responsible for approximately 20% of world greenhouse gas emissions. According to the Intergovernmental Panel on Climate Change deforestation, mainly in tropical areas, could account for up to one-third of total anthropogenic carbon dioxide emissions. But recent calculations suggest that carbon dioxide emissions from deforestation and forest degradation (excluding peatland emissions) contribute about 12% of total anthropogenic carbon dioxide emissions with a range from 6 to 17%. Deforestation causes carbon dioxide to linger in the atmosphere. As carbon dioxide accrues, it produces a layer in the atmosphere that traps radiation from the sun. The radiation converts to heat which causes global warming, which is better known as the greenhouse effect. Plants remove carbon in the form of carbon dioxide from the atmosphere during the process of photosynthesis, but release some carbon dioxide back into the atmosphere during normal respiration. Only when actively growing can a tree or forest remove carbon, by storing it in plant tissues. Both the decay and burning of wood releases much of this stored carbon back to the atmosphere. In order for forests to take up carbon, there must be a net accumulation of wood. One way is for the wood to be harvested and turned into long-lived products, with new young trees replacing them. Deforestation may also cause carbon stores held in soil to be released. Forests can be either sinks or sources depending upon environmental circumstances. Mature forests alternate between being net sinks and net sources of carbon dioxide.

In deforested areas, the land heats up faster and reaches a higher temperature, leading to localized upward motions that enhance the formation of clouds and ultimately produce more rainfall. However, according to the Geophysical Fluid Dynamics Laboratory, the models used to investigate remote responses to tropical deforestation showed a broad but mild temperature increase all through the tropical atmosphere. The model predicted <0.2 °C warming for upper air at 700 mb and 500 mb. However, the model shows no significant changes in other areas besides the Tropics. Though the model showed no significant changes to the climate in areas other than the Tropics, this may not be the case since the model has possible errors and the results are never absolutely definite.

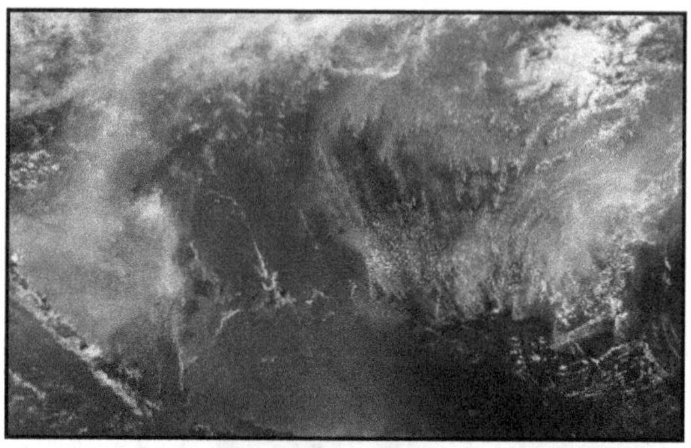

Fires on Borneo and Sumatra, 2006. People use slash-and-burn deforestation to clear land for agriculture.

Reducing emissions from deforestation and forest degradation (REDD) in developing countries has emerged as a new potential to complement ongoing climate policies. The idea consists in providing financial compensations for the reduction of greenhouse gas (GHG) emissions from deforestation and forest degradation".

Rainforests are widely believed by laymen to contribute a significant amount of the world's oxygen, although it is now accepted by scientists that rainforests contribute little net oxygen to the atmosphere and deforestation has only a minor effect on atmospheric oxygen levels. However, the incineration and burning of forest plants to clear land releases large amounts of CO_2, which contributes to global warming. Scientists also state that tropical deforestation releases 1.5 billion tons of carbon each year into the atmosphere.

Hydrological

The water cycle is also affected by deforestation. Trees extract groundwater through their roots and release it into the atmosphere. When part of a forest is removed, the trees no longer transpire this water, resulting in a much drier climate. Deforestation reduces the content of water in the soil and groundwater as well as atmospheric moisture. The dry soil leads to lower water intake for the trees to extract. Deforestation reduces soil cohesion, so that erosion, flooding and landslides ensue.

Shrinking forest cover lessens the landscape's capacity to intercept, retain and transpire precipitation. Instead of trapping precipitation, which then percolates to groundwater systems, deforested areas become sources of surface water runoff, which moves much faster than subsurface flows. Forests return most of the water that falls as precipitation to the atmosphere by transpiration. In contrast, when an area is deforested, almost all precipitation is lost as run-off. That quicker transport of surface water can translate into flash flooding and more localized floods than would occur with the forest cover. Deforestation also contributes to decreased evapotranspiration, which lessens atmospheric moisture which in some cases affects precipitation levels downwind from the deforested area, as water is not recycled to downwind forests, but is lost in runoff and returns directly to the oceans. According to one study, in deforested north and northwest China, the average annual precipitation decreased by one third between the 1950s and the 1980s.

Trees, and plants in general, affect the water cycle significantly:

- Their canopies intercept a proportion of precipitation, which is then evaporated back to the atmosphere (canopy interception).
- Their litter, stems and trunks slow down surface runoff.
- Their roots create macropores – large conduits – in the soil that increase infiltration of water.
- They contribute to terrestrial evaporation and reduce soil moisture via transpiration.

- Their litter and other organic residue change soil properties that affect the capacity of soil to store water.
- Their leaves control the humidity of the atmosphere by transpiring. 99% of the water absorbed by the roots moves up to the leaves and is transpired.

As a result, the presence or absence of trees can change the quantity of water on the surface, in the soil or groundwater, or in the atmosphere. This in turn changes erosion rates and the availability of water for either ecosystem functions or human services.

The forest may have little impact on flooding in the case of large rainfall events, which overwhelm the storage capacity of forest soil if the soils are at or close to saturation.

Tropical rainforests produce about 30% of our planet's fresh water.

Soil

Undisturbed forests have a very low rate of soil loss (erosion), approximately 2 metric tons per square kilometer (6 short tons per square mile). Deforestation generally increases rates of soil loss, by increasing the amount of runoff and reducing the protection of the soil from tree litter. This can be an advantage in excessively leached tropical rain forest soils. Forestry operations themselves also increase erosion through the development of (forest) roads and the use of mechanized equipment.

China's Loess Plateau was cleared of forest millennia ago. Since then it has been eroding, creating dramatic incised valleys, and providing the sediment that gives the Yellow River its yellow color and that causes the flooding of the river in the lower reaches (hence the river's nickname 'China's sorrow').

Removal of trees does not always increase erosion rates. In certain regions of southwest US, shrubs and trees have been encroaching on grassland. The trees themselves enhance the loss of grass between tree canopies. The bare intercanopy areas become highly erodible. The US Forest Service, in Bandelier National Monument for example, is studying how to restore the former ecosystem, and reduce erosion, by removing the trees.

Tree roots bind soil together, and if the soil is sufficiently shallow they act to keep the soil in place by also binding with underlying bedrock. Tree removal on steep slopes with shallow soil thus increases the risk of landslides, which can threaten people living nearby.

Biodiversity

Deforestation on a human scale results in decline in biodiversity, and on a natural global scale is known to cause the extinction of many species. The removal or destruction of areas of forest cover has resulted in a degraded environment with reduced biodiversity. Forests support biodiversity, providing habitat for wildlife; moreover, forests foster medicinal conservation. With forest biotopes being irreplaceable source of new drugs (such as taxol), deforestation can destroy genetic variations (such as crop resistance) irretrievably.

Illegal logging in Madagascar. In 2009, the vast majority of the illegally obtained rosewood was exported to China.

Since the tropical rainforests are the most diverse ecosystems on Earth and about 80% of the world's known biodiversity could be found in tropical rainforests, removal or destruction of significant areas of forest cover has resulted in a degraded environment with reduced biodiversity. A study in Rondônia, Brazil, has shown that deforestation also removes the microbial community which is involved in the recycling of nutrients, the production of clean water and the removal of pollutants.

It has been estimated that we are losing 137 plant, animal and insect species every single day due to rainforest deforestation, which equates to 50,000 species a year. Others state that tropical rainforest deforestation is contributing to the ongoing Holocene mass extinction. The known extinction rates from deforestation rates are very low, approximately 1 species per year from mammals and birds which extrapolates to approximately 23,000 species per year for all species. Predictions have been made that more than 40% of the animal and plant species in Southeast Asia could be wiped out in the 21st century. Such predictions were called into question by 1995 data that show that within regions of Southeast Asia much of the original forest has been converted to monospecific plantations, but that potentially endangered species are few and tree flora remains widespread and stable.

Scientific understanding of the process of extinction is insufficient to accurately make predictions about the impact of deforestation on biodiversity. Most predictions of forestry related biodiversity loss are based on species-area models, with an underlying assumption that as the forest declines species diversity will decline similarly. However, many such models have been proven to be wrong and loss of habitat does not necessarily lead to large scale loss of species. Species-area models are known to overpredict the number of species known to be threatened in areas where actual deforestation is ongoing, and greatly overpredict the number of threatened species that are widespread.

A recent study of the Brazilian Amazon predicts that despite a lack of extinctions thus far, up to 90 percent of predicted extinctions will finally occur in the next 40 years.

Economic Impact

Damage to forests and other aspects of nature could halve living standards for the world's poor and reduce global GDP by about 7% by 2050, a report concluded at the Convention on Biological Diversity (CBD) meeting in Bonn in 2008. Historically, utilization of forest products, including timber and fuel wood, has played a key role in human societies, comparable to the roles of water and cultivable land. Today, developed countries continue to utilize timber for building houses, and wood pulp for paper. In developing countries almost three billion people rely on wood for heating and cooking.

The forest products industry is a large part of the economy in both developed and developing countries. Short-term economic gains made by conversion of forest to agriculture, or over-exploitation of wood products, typically leads to loss of long-term income and long-term biological productivity. West Africa, Madagascar, Southeast Asia and many other regions have experienced lower revenue because of declining timber harvests. Illegal logging causes billions of dollars of losses to national economies annually.

The new procedures to get amounts of wood are causing more harm to the economy and overpower the amount of money spent by people employed in logging. According to a study, "in most areas studied, the various ventures that prompted deforestation rarely generated more than US$5 for every ton of carbon they released and frequently returned far less than US$1". The price on the European market for an offset tied to a one-ton reduction in carbon is 23 euro (about US$35).

Rapidly growing economies also have an effect on deforestation. Most pressure will come from the world's developing countries, which have the fastest-growing populations and most rapid economic (industrial) growth. In 1995, economic growth in developing countries reached nearly 6%, compared with the 2% growth rate for developed countries." As our human population grows, new homes, communities, and expansions of cities will occur. Connecting all of the new expansions will be roads, a very important part in our daily life. Rural roads promote economic development but also facilitate deforestation. About 90% of the deforestation has occurred within 100 km of roads in most parts of the Amazon.

The European Union is one of the largest importer of products made from illegal deforestation.

Forest Transition Theory

The forest area change may follow a pattern suggested by the forest transition (FT) theory, whereby at early stages in its development a country is characterized by high forest cover and low deforestation rates (HFLD countries).

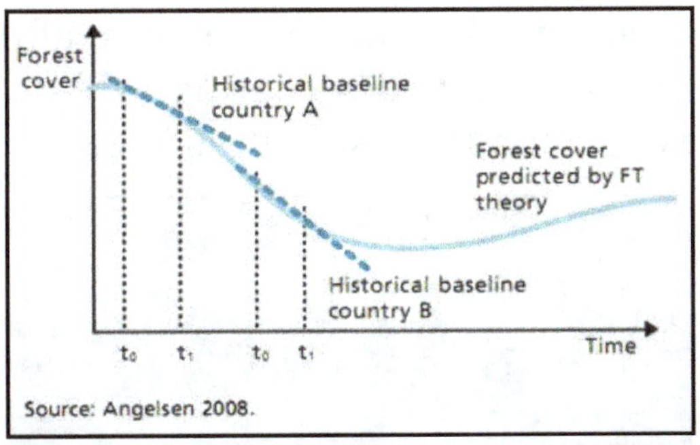

The forest transition and historical baselines.

Then deforestation rates accelerate (HFHD, high forest cover – high deforestation rate), and forest cover is reduced (LFHD, low forest cover – high deforestation rate), before the deforestation rate slows (LFLD, low forest cover – low deforestation rate), after which forest cover stabilizes and eventually starts recovering. FT is not a "law of nature," and the pattern is influenced by national context (for example, human population density, stage of development, structure of the economy), global economic forces, and government policies. A country may reach very low levels of forest cover before it stabilizes, or it might through good policies be able to "bridge" the forest transition.

FT depicts a broad trend, and an extrapolation of historical rates therefore tends to underestimate future BAU deforestation for counties at the early stages in the transition (HFLD), while it tends to overestimate BAU deforestation for countries at the later stages (LFHD and LFLD).

Countries with high forest cover can be expected to be at early stages of the FT. GDP per capita captures the stage in a country's economic development, which is linked to the pattern of natural resource use, including forests. The choice of forest cover and GDP per capita also fits well with the two key scenarios in the FT:

(i) a forest scarcity path, where forest scarcity triggers forces (for example, higher prices of forest products) that lead to forest cover stabilization;

(ii) an economic development path, where new and better off-farm employment opportunities associated with economic growth (= increasing GDP per capita) reduce profitability of frontier agriculture and slows deforestation.

Historical Causes

Prehistory

The Carboniferous Rainforest Collapse was an event that occurred 300 million years ago. Climate change devastated tropical rainforests causing the extinction of many plant and animal species. The change was abrupt, specifically, at this time climate became cooler and drier, conditions that are not favourable to the growth of rainforests and much of the biodiversity within them. Rainforests were fragmented forming shrinking 'islands' further and further apart. This sudden collapse affected several large groups, effects on amphibians were particularly devastating, while reptiles fared better, being ecologically adapted to the drier conditions that followed.

An array of Neolithic artifacts, including bracelets, axe heads, chisels, and polishing tools.

Rainforests once covered 14% of the earth's land surface; now they cover a mere 6% and experts estimate that the last remaining rainforests could be consumed in less than 40 years. Small scale deforestation was practiced by some societies for tens of thousands of years before the beginnings of civilization. The first evidence of deforestation appears in the Mesolithic period. It was probably used to convert closed forests into more open ecosystems favourable to game animals. With the advent of agriculture, larger areas began to be deforested, and fire became the prime tool to clear land for crops. In Europe there is little solid evidence before 7000 BC. Mesolithic foragers used fire to create openings for red deer and wild boar. In Great Britain, shade-tolerant species such as oak and ash are replaced in the pollen record by hazels, brambles, grasses and nettles. Removal of the forests led to decreased transpiration, resulting in the formation of upland peat bogs. Widespread decrease in elm pollen across Europe between 8400–8300 BC and 7200–7000 BC, starting in southern Europe and gradually moving north to Great Britain, may represent land clearing by fire at the onset of Neolithic agriculture.

The Neolithic period saw extensive deforestation for farming land. Stone axes were being made from about 3000 BC not just from flint, but from a wide variety of hard rocks from across Britain and North America as well. They include the noted Langdale axe industry in the English Lake District, quarries developed at Penmaenmawr in North Wales and numerous other locations. Rough-outs were made locally near the quarries, and some were polished locally to give a fine finish. This step not only increased the mechanical strength of the axe, but also made penetration of wood easier. Flint was still used from sources such as Grimes Graves but from many other mines across Europe.

Evidence of deforestation has been found in Minoan Crete; for example the environs of the Palace of Knossos were severely deforested in the Bronze Age.

Pre-industrial History

Throughout most of history, humans were hunter gatherers who hunted within forests. In most areas, such as the Amazon, the tropics, Central America, and the Caribbean, only after shortages of wood and other forest products occur are policies implemented to ensure forest resources are used in a sustainable manner.

In ancient Greece, Tjeered van Andel and co-writers summarized three regional studies of historic erosion and alluviation and found that, wherever adequate evidence exists, a major phase of erosion follows, by about 500-1,000 years the introduction of farming in the various regions of Greece, ranging from the later Neolithic to the Early Bronze Age. The thousand years following the mid-first millennium BC saw serious, intermittent pulses of soil erosion in numerous places. The historic silting of ports along the southern coasts of Asia Minor (*e.g.* Clarus, and the examples of Ephesus, Priene and Miletus, where harbors had to be abandoned because of the silt deposited by the Meander) and in coastal Syria during the last centuries BC.

Easter Island

Easter Island has suffered from heavy soil erosion in recent centuries, aggravated by agriculture and deforestation. Jared Diamond gives an extensive look into the collapse of the ancient Easter Islanders in his book *Collapse*. The disappearance of the island's trees seems to coincide with a decline of its civilization around the 17th and 18th century. He attributed the collapse to deforestation and over-exploitation of all resources.

The famous silting up of the harbor for Bruges, which moved port commerce to Antwerp, also followed a period of increased settlement growth (and apparently of deforestation) in the upper river basins. In early medieval Riez in upper Provence, alluvial silt from two small rivers raised the riverbeds and widened the floodplain, which slowly buried the Roman settlement in alluvium and gradually moved new construction to higher ground; concurrently the headwater valleys above Riez were being opened to pasturage.

A typical progress trap was that cities were often built in a forested area, which would provide wood for some industry (for example, construction, shipbuilding, pottery). When deforestation occurs without proper replanting, however; local wood supplies become difficult to obtain near enough to remain competitive, leading to the city's abandonment, as happened repeatedly in Ancient Asia Minor. Because of fuel needs, mining and metallurgy often led to deforestation and city abandonment.

With most of the population remaining active in (or indirectly dependent on) the agricultural sector, the main pressure in most areas remained land clearing for crop and cattle farming. Enough wild green was usually left standing (and partially used, for example, to collect firewood, timber and fruits, or to graze pigs) for wildlife to remain viable. The elite's (nobility and higher clergy) protection of their own hunting privileges and game often protected significant woodlands.

Major parts in the spread (and thus more durable growth) of the population were played by monastical 'pioneering' (especially by the Benedictine and Commercial orders) and some feudal lords' recruiting farmers to settle (and become tax payers) by offering relatively good legal and fiscal conditions. Even when speculators sought to encourage towns, settlers needed an agricultural belt around or sometimes within defensive walls. When populations were quickly decreased by causes such as the Black Death or devastating warfare (for example, Genghis Khan's Mongol hordes in eastern and central Europe, Thirty Years' War in Germany), this could lead to settlements being abandoned. The land was reclaimed by nature, but the secondary forests usually lacked the original biodiversity.

Deforestation of Brazil's Atlantic Forest c.1820-1825.

From 1100 to 1500 AD, significant deforestation took place in Western Europe as a result of the expanding human population. The large-scale building of wooden sailing ships by European (coastal) naval owners since the 15th century for exploration, colonisation, slave trade–and other trade on the high seas consumed many forest resources. Piracy also contributed to the over harvesting of forests, as in Spain. This led to a weakening of the domestic economy after Columbus' discovery of America, as the economy became dependent on colonial activities (plundering, mining, cattle, plantations, trade, etc.)

In *Changes in the Land* (1983), William Cronon analyzed and documented 17th-century English colonists' reports of increased seasonal flooding in New England during the period when new settlers initially cleared the forests for agriculture. They believed flooding was linked to widespread forest clearing upstream.

The massive use of charcoal on an industrial scale in Early Modern Europe was a new type of consumption of western forests; even in Stuart England, the relatively primitive production of charcoal has already reached an impressive level. Stuart England was so widely deforested that it depended on the Baltic trade for ship timbers, and looked to the untapped forests of New England to supply the need. Each of Nelson's Royal Navy war ships at Trafalgar (1805) required 6,000 mature oaks for its construction. In France, Colbert planted oak forests to supply the French navy in the future. When the oak plantations matured in the mid-19th century, the masts were no longer required because shipping had changed.

Norman F. Cantor's summary of the effects of late medieval deforestation applies equally well to Early Modern Europe:

Europeans had lived in the midst of vast forests throughout the earlier medieval centuries. After 1250 they became so skilled at deforestation that by 1500 they were running short of wood for heating and cooking. They were faced with a nutritional decline because of the elimination of the generous supply of wild game that had inhabited the now-disappearing forests, which throughout medieval times had provided the staple of their carnivorous high-protein diet. By 1500 Europe was on the edge of a fuel and nutritional disaster [from] which it was saved in the sixteenth century only by the burning of soft coal and the cultivation of potatoes and maize.

Industrial Era

In the 19th century, introduction of steamboats in the United States was the cause of deforestation of banks of major rivers, such as the Mississippi River, with increased and more severe flooding one of the environmental results. The steamboat crews cut wood every day from the riverbanks to fuel the steam engines. Between St. Louis and the confluence with the Ohio River to the south, the Mississippi became more wide and shallow, and changed its channel laterally. Attempts to improve navigation by the use of snag pullers often resulted in crews' clearing large trees 100 to 200 feet (61 m) back from the banks. Several French colonial towns of the Illinois Country, such as Kaskaskia, Cahokia and St. Philippe, Illinois were flooded and abandoned in the late 19th century, with a loss to the cultural record of their archeology.

The wholescale clearance of woodland to create agricultural land can be seen in many parts of the world, such as the Central forest-grasslands transition and other areas of the Great Plains of the United States. Specific parallels are seen in the 20th-century deforestation occurring in many developing nations.

Rates of Deforestation

Global deforestation sharply accelerated around 1852. It has been estimated that about half of the Earth's mature tropical forests—between 7.5 million and 8 million km^2 (2.9 million to 3 million sq mi) of the original 15 million to 16 million km^2 (5.8 million to 6.2 million sq mi) that until 1947 covered the planet—have now been destroyed. Some scientists have predicted that unless significant measures (such as seeking out and protecting old growth forests that have not been disturbed) are taken on a worldwide basis, by 2030 there will only be 10% remaining, with another 10% in a degraded condition. 80% will have been lost, and with them hundreds of thousands of irreplaceable species. Some cartographers have attempted to illustrate the sheer scale of deforestation by country using a cartogram.

Slash-and-burn farming in the state of Rondônia, western Brazil.

Estimates vary widely as to the extent of tropical deforestation. Scientists estimate that one fifth of the world's tropical rainforest was destroyed between 1960 and 1990. They claim that that rainforests 60 years ago covered 14% of the world's land surface, now only cover 5–7%, and that all tropical forests will be gone by the middle of the 21st century.

A 2002 analysis of satellite imagery suggested that the rate of deforestation in the humid tropics (approximately 5.8 million hectares per year) was roughly 23% lower than the most commonly quoted rates. Conversely, a newer analysis of satellite images reveals that deforestation of the Amazon rainforest is twice as fast as scientists previously estimated.

Some have argued that deforestation trends may follow a Kuznets curve, which if true would nonetheless fail to eliminate the risk of irreversible loss of non-economic forest values (for example, the extinction of species).

Satellite image of Haiti's border with the Dominican Republic (right) shows the amount of deforestation on the Haitian side.

A 2005 report by the United Nations Food and Agriculture Organization (FAO) estimates that although the Earth's total forest area continues to decrease at about 13 million hectares per year, the global rate of deforestation has recently been slowing. Still others claim that rainforests are being destroyed at an ever-quickening pace. The London-based Rainforest Foundation notes that "the UN figure is based on a definition of forest as being an area with as little as 10% actual tree cover, which would therefore include areas that are actually savannah-like ecosystems and badly damaged forests." Other critics of the FAO data point out that they do not distinguish between forest types, and that they are based largely on reporting from forestry departments of individual countries, which do not take into account unofficial activities like illegal logging.

Despite these uncertainties, there is agreement that destruction of rainforests remains a significant environmental problem. Up to 90% of West Africa's coastal rainforests have disappeared since 1900. In South Asia, about 88% of the rainforests have been lost. Much of what remains of the world's rainforests is in the Amazon basin, where the Amazon Rainforest covers approximately 4 million square kilometres. The regions with the highest tropical deforestation rate between 2000 and 2005 were Central America—which lost 1.3% of its forests each year—and tropical Asia. In Central America, two-thirds of lowland tropical forests have been turned into pasture since 1950 and 40% of all the rainforests have been lost in the last 40 years. Brazil has lost 90–95% of its Mata Atlântica forest. Paraguay was losing its natural semi humid forests in the country's western regions at a rate of 15.000 hectares at a randomly studied 2-month period in 2010, Paraguay's parliament refused in 2009 to pass a law that would have stopped cutting of natural forests altogether.

Deforestation around Pakke Tiger Reserve, India.

Madagascar has lost 90% of its eastern rainforests. As of 2007, less than 1% of Haiti's forests remained. Mexico, India, the Philippines, Indonesia, Thailand, Burma, Malaysia, Bangladesh, China, Sri Lanka, Laos, Nigeria, the Democratic Republic of the Congo, Liberia, Guinea, Ghana and the Ivory Coast, have lost large areas of their rainforest. Several countries, notably Brazil, have declared their deforestation a national emergency. The World Wildlife Fund's ecoregion project catalogues habitat types throughout the world, including habitat loss such as deforestation, showing for example that even in the rich forests of parts of Canada such as the Mid-Continental Canadian forests of the prairie provinces half of the forest cover has been lost or altered.

Regions

Rates of deforestation vary around the world.

In 2011 Conservation International listed the top 10 most endangered forests, characterized by having all lost 90% or more of their original habitat, and each harboring at least 1500 endemic plant species (species found nowhere else in the world).

Top 10 Most Endangered Forests 2011				
Endangered forest	Region	Remaining habitat	Predominate vegetation type	Notes
Indo-Burma	Asia-Pacific	5%	Tropical and subtropical moist broadleaf forests	Rivers, floodplain wetlands, mangrove forests. Burma, Thailand, Laos, Vietnam, Cambodia, India.
New Caledonia	Asia-Pacific	5%	Tropical and subtropical moist broadleaf forests	See note for region covered.
Sundaland	Asia-Pacific	7%	Tropical and subtropical moist broadleaf forests	Western half of the Indo-Malayan archipelago including southern Borneo and Sumatra.
Philippines	Asia-Pacific	7%	Tropical and subtropical moist broadleaf forests	Forests over the entire country including 7,100 islands.
Atlantic Forest	South America	8%	Tropical and subtropical moist broadleaf forests	Forests along Brazil's Atlantic coast, extends to parts of Paraguay, Argentina and Uruguay.
Mountains of Southwest China	Asia-Pacific	8%	Temperate coniferous forest	See note for region covered.
California Floristic Province	North America	10%	Tropical and subtropical dry broadleaf forests	See note for region covered.
Coastal Forests of Eastern Africa	Africa	10%	Tropical and subtropical moist broadleaf forests	Mozambique, Tanzania, Kenya, Somalia.
Madagascar & Indian Ocean Islands	Africa	10%	Tropical and subtropical moist broadleaf forests	Madagascar, Mauritius, Reunion, Seychelles, Comoros.
Eastern Afromontane	Africa	11%	Tropical and subtropical moist broadleaf forests Montane grasslands and shrublands	Forests scattered along the eastern edge of Africa, from Saudi Arabia in the north to Zimbabwe in the south.

Control

Reducing Emissions

Main international organizations including the United Nations and the World Bank, have begun to develop programs aimed at curbing deforestation. The blanket term Reducing Emissions from Deforestation and Forest Degradation (REDD) describes these sorts of programs, which use direct monetary or other incentives to encourage developing countries to limit and/or roll back deforestation. Funding has been an issue, but at the UN Framework Convention on Climate Change (UNFCCC) Conference of the Parties-15 (COP-15) in Copenhagen in December 2009, an accord was reached with a collective commitment by developed countries for new and additional resources, including forestry and investments through international institutions, that will approach USD 30 billion for the period 2010–2012. Significant work is underway on tools for use in monitoring developing country adherence to their agreed REDD targets. These tools, which rely on remote forest monitoring using satellite imagery and other data sources, include the Center for Global Development's FORMA (Forest Monitoring for Action) initiative and the Group on Earth Observations' Forest Carbon Tracking Portal. Methodological guidance for forest monitoring was also emphasized at COP-15. The environmental organization Avoided Deforestation Partners leads the campaign for development of REDD through funding from the U.S. government. In 2014, the Food and Agriculture Organization of the United Nations and partners launched Open Foris - a set of open-source software tools that assist countries in gathering, producing and disseminating information on the state of forest resources. The tools support the inventory lifecycle, from needs assessment, design, planning, field data collection and management, estimation analysis, and dissemination. Remote sensing image processing tools are included, as well as tools for international reporting for Reducing emissions from deforestation and forest degradation (REDD) and MRV and FAO's Global Forest Resource Assessments.

In evaluating implications of overall emissions reductions, countries of greatest concern are those categorized as High Forest Cover with High Rates of Deforestation (HFHD) and Low Forest Cover with High Rates of Deforestation (LFHD). Afghanistan, Benin, Botswana, Burma, Burundi, Cameroon, Chad, Ecuador, El Salvador, Ethiopia, Ghana, Guatemala, Guinea, Haiti, Honduras, Indonesia, Liberia, Malawi, Mali, Mauritania, Mongolia, Namibia, Nepal, Nicaragua, Niger, Nigeria, Pakistan, Paraguay, Philippines, Senegal, Sierra Leone, Sri Lanka, Sudan, Togo, Uganda, United Republic of Tanzania, Zimbabwe are listed as having Low Forest Cover with High Rates of Deforestation (LFHD). Brazil, Cambodia, Democratic Peoples Republic of Korea, Equatorial Guinea, Malaysia, Solomon Islands, Timor-Leste, Venezuela, Zambia are listed as High Forest Cover with High Rates of Deforestation (HFHD).

Payments for Conserving Forests

In Bolivia, deforestation in upper river basins has caused environmental problems, including soil erosion and declining water quality. An innovative project to try and remedy this situation involves landholders in upstream areas being paid by downstream water users to conserve forests. The landholders receive US$20 to conserve the trees, avoid polluting livestock practices, and enhance the biodiversity and forest carbon on their land. They also receive US$30, which purchases a beehive, to compensate for conservation for two hectares of water-sustaining forest for five years. Honey revenue per hectare of forest is US$5 per year, so within five years, the landholder has sold US$50 of honey. The project is being conducted by Fundación Natura Bolivia and Rare Conservation, with support from the Climate & Development Knowledge Network.

Land Rights

Transferring land rights to indigenous inhabitants is argued to efficiently conserve forests.

Transferring rights over land from public domain to its indigenous inhabitants is argued to be a cost effective strategy to conserve forests. This includes the protection of such rights entitled in existing laws, such as India's Forest Rights Act. The transferring of such rights in China, perhaps the largest land reform in modern times, has been argued to have increased forest cover. In Brazil, forested areas given tenure to indigenous groups have even lower rates of clearing than national parks.

Farming

New methods are being developed to farm more intensively, such as high-yield hybrid crops, greenhouse, autonomous building gardens, and hydroponics. These methods are often dependent on chemical inputs to maintain necessary yields. In cyclic agriculture, cattle are grazed on farm land that is resting and rejuvenating. Cyclic agriculture actually increases the fertility of the soil. Intensive farming can also decrease soil nutrients by consuming at an accelerated rate the trace minerals needed for crop growth.The most promising approach, however, is the concept of food forests in permaculture, which consists of agroforestal systems carefully designed to mimic natural forests, with an emphasis on plant and animal species of interest for food, timber and other uses. These systems have low dependence on fossil fuels and agro-chemicals, are highly self-maintaining, highly productive, and with strong positive impact on soil and water quality, and biodiversity.

Monitoring Deforestation

There are multiple methods that are appropriate and reliable for reducing and monitoring deforestation. One method is the "visual interpretation of aerial photos or satellite imagery that is labor-intensive but does not require high-level training in computer image processing or extensive computational resources". Another method includes hot-spot analysis (that is, locations of rapid change) using expert opinion or coarse resolution satellite data to identify locations for detailed digital analysis with high resolution satellite images. Deforestation is typically assessed by quantifying the amount of area deforested, measured at the present time. From an environmental point of view, quantifying the damage and its possible consequences is a more important task, while conservation efforts are more focused on forested land protection and development of land-use alternatives to avoid continued deforestation. Deforestation rate and total area deforested, have been widely used for monitoring deforestation in many regions, including the Brazilian Amazon deforestation monitoring by INPE.

Forest Management

Efforts to stop or slow deforestation have been attempted for many centuries because it has long been known that deforestation can cause environmental damage sufficient in some cases to cause societies to collapse. In Tonga, paramount rulers developed policies designed to prevent conflicts between short-term gains from converting forest to farmland and long-term problems forest loss would cause, while during the 17th and 18th centuries in Tokugawa, Japan, the shoguns developed a highly sophisticated system of long-term planning to stop and even reverse deforestation of the preceding centuries through substituting timber by other products and more efficient use of land that had been farmed for many centuries. In 16th-century Germany, landowners also developed silviculture to deal with the problem of deforestation. However, these policies tend to be limited to environments with *good rainfall, no dry season* and *very young soils* (through volcanism or glaciation). This is because on older and less fertile soils trees grow too slowly for silviculture to be economic, whilst in areas with a strong dry season there is always a risk of forest fires destroying a tree crop before it matures.

In the areas where "slash-and-burn" is practiced, switching to "slash-and-char" would prevent the rapid deforestation and subsequent degradation of soils. The biochar thus created, given back to the soil, is not only a durable carbon sequestration method, but it also is an extremely beneficial amendment to the soil. Mixed with biomass it brings the creation of terra preta, one of the richest soils on the planet and the only one known to regenerate itself.

Sustainable Practices

Certification, as provided by global certification systems such as Programme for the Endorsement of Forest Certification and Forest Stewardship Council, contributes to tackling deforestation by creating market demand for timber from sustainably managed forests. According to the United Nations Food and Agriculture Organization (FAO), "A major condition for the adoption of sustainable forest management is a demand for products that are produced sustainably and consumer willingness to pay for the higher costs entailed. Certification represents a shift from regulatory approaches to market incentives to promote sustainable forest management. By promoting the positive attributes of forest products from sustainably managed forests, certification focuses on the demand side of environmental conservation." Rainforest Rescue argues that the standards of organizations like FSC are too closely connected to timber industry interests and therefore do not guarantee environmentally and socially responsible forest management. In reality, monitoring systems are inadequate and various cases of fraud have been documented worldwide.

Bamboo is advocated as a more sustainable alternative for cutting down wood for fuel.

Some nations have taken steps to help increase the amount of trees on Earth. In 1981, China created National Tree Planting Day Forest and forest coverage had now reached 16.55% of China's land mass, as against only 12% two decades ago.

Using fuel from bamboo rather than wood results in cleaner burning, and since bamboo matures much faster than wood, deforestation is reduced as supply can be replenished faster.

Reforestation

In many parts of the world, especially in East Asian countries, reforestation and afforestation are increasing the area of forested lands. The amount of woodland has increased in 22 of the world's 50 most forested nations. Asia as a whole gained 1 million hectares of forest between 2000 and 2005. Tropical forest in El Salvador expanded more than 20% between 1992 and 2001. Based on these trends, one study projects that global forest will increase by 10%—an area the size of India—by 2050.

In the People's Republic of China, where large scale destruction of forests has occurred, the government has in the past required that every able-bodied citizen between the ages of 11 and 60 plant three to five trees per year or do the equivalent amount of work in other forest services. The government claims that at least 1 billion trees have been planted in China every year since 1982. This is no longer required today, but March 12 of every year in China is the Planting Holiday. Also, it has introduced the Green Wall of China project, which aims to halt the expansion of the Gobi desert through the planting of trees. However, due to the large percentage of trees dying off after planting (up to 75%), the project is not very successful. There has been a 47-million-hectare increase in forest area in China since the 1970s. The total number of trees amounted to be about 35 billion and 4.55% of China's land mass increased in forest coverage. The forest coverage was 12% two decades ago and now is 16.55%.

An ambitious proposal for China is the Aerially Delivered Re-forestation and Erosion Control System and the proposed Sahara Forest Project coupled with the Seawater Greenhouse.

In Western countries, increasing consumer demand for wood products that have been produced and harvested in a sustainable manner is causing forest landowners and forest industries to become increasingly accountable for their forest management and timber harvesting practices.

The Arbor Day Foundation's Rain Forest Rescue program is a charity that helps to prevent deforestation. The charity uses donated money to buy up and preserve rainforest land before the lumber companies can buy it. The Arbor Day Foundation then protects the land from deforestation. This also locks in the way of life of the primitive tribes living on the forest land. Organizations such as Community Forestry International, Cool Earth, The Nature Conservancy, World Wide Fund for Nature, Conservation International, African Conservation Foundation and Greenpeace also focus on preserving forest habitats. Greenpeace in particular has also mapped out the forests that are still intact and published this information on the internet. World Resources Institute in turn has made a simpler thematic map showing the amount of forests present just before the age of man (8000 years ago) and the current (reduced) levels of forest. These maps mark the amount of afforestation required to repair the damage caused by people.

Forest Plantations

To meet the world's demand for wood, it has been suggested by forestry writers Botkins and Sedjo that high-yielding forest plantations are suitable. It has been calculated that plantations yielding 10 cubic meters per hectare annually could supply all the timber required for international trade on 5% of the world's existing forestland. By contrast, natural forests produce about 1–2 cubic meters per hectare; therefore, 5–10 times more forestland would be required to meet demand. Forester Chad Oliver has suggested a forest mosaic with high-yield forest lands interspersed with conservation land.

In the country of Senegal, on the western coast of Africa, a movement headed by youths has helped to plant over 6 million mangrove trees. The trees will protect local villages from storm damages and will provide a habitat for local wildlife. The project started in 2008, and already the Senegalese government has been asked to establish rules and regulations that would protect the new mangrove forests.

Military Context

American Sherman tanks knocked out by Japanese artillery on Okinawa.

While the preponderance of deforestation is due to demands for agricultural and urban use for the human population, there are some examples of military causes. One example of deliberate deforestation is that which took place in the U.S. zone of occupation in Germany after World War II. Before the onset of the Cold War, defeated Germany was still considered a potential future threat rather than potential future ally. To address this threat, attempts were made to lower German industrial potential, of which forests were deemed an element. Sources in the U.S. government admitted that the purpose of this was that the "ultimate destruction of the war potential of German forests." As a consequence of the practice of clear-felling, deforestation resulted which could "be replaced only by long forestry development over perhaps a century."

Deforestation can also be one consequence of war. For example, in the 1945 Battle of Okinawa, bombardment and other combat operations reduced the lush tropical landscape into "a vast field of mud, lead, decay and maggots". Deforestation can also be an intentional tactic of military forces. Defoliants (like Agent Orange or others) was used by the British in the Malayan Emergency, and by the United States in the Korean War and Vietnam War.

Public Health Context

Deforestation eliminates a great number of species of plants and animals which also often results in an increase in disease. Loss of native species allows new species to come to dominance. Often the destruction of predatory species can result in an increase in rodent populations. These are known to carry plagues. Additionally, erosion can produce pools of stagnant water that are perfect breeding grounds for mosquitos, well known vectors of malaria, yellow fever, nipah virus, and more. Deforestation can also create a path for non-native species to flourish such as certain types of snails, which have been correlated with an increase in schistosomiasis cases.

Deforestation is occurring all over the world and has been coupled with an increase in the occurrence of disease outbreaks. In Malaysia, thousands of acres of forest have been cleared for pig farms. This has resulted in an increase in the zoonosis the Nipah virus. In Kenya, deforestation has led to an increase in malaria cases which is now the leading cause of morbidity and mortality the country.

Another pathway through which deforestation affects disease is the relocation and dispersion of disease-carrying hosts. This disease emergence pathway can be called "range expansion," whereby the host's range (and thereby the range of pathogens) expands to new geographic areas. Through deforestation, hosts and reservoir species are forced into neighboring habitats. Accompanying the reservoir species are pathogens that have the ability to find new hosts in previously unexposed regions. As these pathogens and species come into closer contact with humans, they are infected both directly and indirectly.

A catastrophic example of range expansion is the 1998 outbreak of Nipah Virus in Malaysia. For a number of years, deforestation, drought, and subsequent fires led to a dramatic geographic shift and density of fruit bats, a reservoir for Nipah virus. Deforestation reduced the available fruiting trees in the bats' habitat, and they encroached on surrounding orchards which also happened to be the location of a large number of pigsties. The bats, through proximity spread the Nipah to pigs. While the virus infected the pigs, mortality was much lower than among humans, making the pigs a virulent host leading to the transmission of the virus to humans. This resulted in 265 reported cases of encephalitis, of which 105 resulted in death. This example provides an important lesson for the impact deforestation can have on human health.

Another example of range expansion due to deforestation and other anthropogenic habitat impacts includes the Capybara rodent in Paraguay. This rodent is the host of a number of zoonotic diseases and, while there has not yet been a human-borne outbreak due to the movement of this rodent into new regions, it offers an example of how habitat destruction through deforestation and subsequent movements of species is occurring regularly.

A now well-developed theory is that the spread of HIV it is at least partially due deforestation. Rising populations created a food demand and with deforestation opening up new areas of the forest the hunters harvested a great deal of primate bushmeat, which is believed to be the origin of HIV.

Illegal Logging

Illegal logging is the harvest, transportation, purchase or sale of timber in violation of laws. The harvesting procedure itself may be illegal, including using corrupt means to gain access to forests; extraction without permission or from a protected area; the cutting of protected species; or the extraction of timber in excess of agreed limits.

Illegalities may also occur during transport, such as illegal processing and export; fraudulent declaration to customs; the avoidance of taxes and other charges, and fraudulent certification.

Overview

Illegal logging is a pervasive problem, causing enormous damage to forests, local communities and to the economies of producer countries. Despite the economic importance of trade in timber and forest products, major international timber consumer countries, such as the EU, have no legal means to halt the import of illegally sourced forest products, because the identification of illegally logged or traded timber is technically difficult. Therefore, a legal basis for normative acts against timber imports or other products manufactured out of illegal wood is missing. Scientific methods to pinpoint the geographic origin of timber are currently under development. Possible actions to restrict imports cannot meet with WTO regulations of non-discrimination. They must instead be arranged in bilateral agreements. TRAFFIC, the wildlife trade monitoring network strives to monitor the illegal trade of timber and provide expertise in policy and legal reviews.

Scale

It is estimated that illegal logging in public lands alone causes losses in assets and revenue in excess of 10 billion USD annually. Although exact figures are difficult to state, given the illegal nature of the activity, decant estimates show that most of the logging, more than half, that is done in the world is illegal most especially in open and vulnerable areas – the Amazon Basin, Central Africa, Southeast Asia, the Russian Federation – is illegal.

Available figures and estimations must be dealt with cautiously. Governments may tend to underestimate the situation. High estimates of illegal logging may constitute an embarrassment as these hint at ineffective enforcement of legislation or, even worse, bribery and corruption. On the other hand, environmental NGOs publish alarming figures to raise awareness and to emphasise the need for stricter conservation measures. For companies of the forest sector, publications providing high estimations can be regarded as potentially threatening for their reputation and their market perspective, including the competitiveness of wood in comparison to other materials. However, for many countries, NGOs are the only source of information apart from state institutions that probably clearly underestimate the situation. For example, the Republic of Estonia calculated an amount of 1% illegally harvested timber in 2003, whereas it is estimated to reach a maximum of 50% by the ENGO "Estonian Green Movement". In Latvia, the situation is comparable; anecdotal evidence points towards 25% of logging being illegal.

Consequences

Illegal logging contributes to deforestation and by extension global warming, causes loss of biodiversity and undermines the rule of law. These illegal activities undermine responsible forest management, encourage corruption and tax evasion and reduce the income of the producer countries, further limiting the resources producer countries can invest in sustainable development. Illegal logging has serious economic and social implications for the poor and disadvantaged with millions of dollars worth of timber revenue being lost each year.

Illegal logging continues in Thailand. This photograph was taken from the roadside in Mae Wang District, Chiang Mai Province, in March 2011.

Furthermore, the illegal trade of forest resources undermines international security, and is frequently associated with corruption, money laundering, organized crime, human rights abuses and, in some cases, violent conflict. In the forestry sector, cheap imports of illegal timber and forest products, together with the non-compliance of some economic players with basic social and environmental standards, destabilise international markets. This unfair competition affects those European companies, especially the small and medium-sized companies that are behaving responsibly and ready to play by fair rules.

Statistics on Illegal Logging

- A joint UK-Indonesian study of the timber industry in Indonesia in 1998 suggested that about 40% of throughput was illegal, with a value in excess of $365 million. More recent estimates, comparing legal harvesting against known domestic consumption plus exports, suggest that 88% of logging in the country is illegal in some way. Malaysia is the key transit country for illegal wood products from Indonesia.
- In Brazil, 80% of logging in the Amazon violates government controls. At the core of illegal logging is widespread corruption. Often referred to as 'green gold', mahogany can fetch over US$1,600 m-3. Illegal mahogany opens the door for illegal logging of other species, and for widespread exploitation of the Brazilian Amazon. Recent Greenpeace investigations in the Brazilian state of Pará reveal just how deeply rooted the problem remains. No reliable legal chain of custody exists for mahogany, and the key players in its trade are ruthless.
- The World Bank estimates that 80% of logging operations are illegal in Bolivia and 42% in Colombia, 10 while in Peru, illegal logging equals 80% of all activities.

- Research carried out by WWF International in 2002 shows that in Africa, rates of illegal logging vary from 50% for Cameroon and Equatorial Guinea to 70% in Gabon and 80% in Liberia – where revenues from the timber industry also fuelled the civil war.
- WWF estimates that illegal logging in Russia is at least 20%, reaching up to 50% in its far eastern regions.
- A 2012 joint study by the United Nations Environment Programme and Interpol states that illegal logging accounts for up to 30% of the global logging trade and contributes to more than 50% of tropical deforestation in Central Africa, the Amazon Basin and South East Asia.
- Between 50% and 90% of logging from the key countries in these regions is being carried out by organised criminal entities.

Political Processes

East Asia

Signs of illegal timber poaching on the boundary of the protected area around the Cagua Volcano, Cagayan, Philippines.

The East Asia Forest Law Enforcement and Governance (EA FLEG) Ministerial Conference took place in Bali in September 2001. The Conference brought together nearly 150 participants from 20 countries, representing government, international organizations, NGOs, and the private sector. The event was co-hosted by the World Bank and the Government of Indonesia.

The meeting included detailed technical discussions of forest law enforcement in relation to governance, forest policy and forest management as well as ministerial engagement.

The Conference's primary aims were to share analysis on forest law enforcement; explore priority issues of forest law enforcement, including illegal logging in the East Asia region, among senior officials from forest and related ministries, NGOs and industry representatives; and commit to action at the national and regional level.

European Union

In May 2003, the European Commission presented the EU Forest Law Enforcement, Governance and Trade Action Plan (EU FLEGT). This marked the beginning of a long process by which the EU aims to develop and implement measures to address illegal logging and related trade. The primary means of implementing the Plan is through Voluntary Partnership Agreements with timber producing countries. The EU Timber Regulation (EUTR) was adopted in 2010 and went into effect 3 March 2013.

- It prohibits the placing on the EU market for the first time of illegally harvested timber and products derived from such timber;
- It requires EU traders who place timber products on the EU market for the first time to exercise 'due diligence';
- Once on the market, the timber and timber products may be sold on and/or transformed before they reach the final consumer;
- To facilitate the traceability of timber products economic operators in this part of the supply chain (referred to as traders in the regulation) have an obligation to keep records of their suppliers and customers.

A Greenpeace investigation published in May 2014 demonstrates that EU Timber Regulation is ineffective if fraudulent paperwork is accepted at face value and there is not sufficient enforcement by EU authorities.

Africa

The *Africa Forest Law Enforcement and Governance* (AFLEG) Ministerial Conference was held in Yaoundé, Cameroon in October 2003. The meeting drew together ministers and stakeholders from Africa, Europe and North America to consider how partnerships between producers, consumers, donors, civil society and the private sector could address illegal forest exploitation and associated trade in Africa.

The AFLEG conference, the second regional forest law enforcement and governance meeting after East Asia, resulted in endorsement of a ministerial declaration and action plan as well as a variety of informal implementation initiatives.

In 2014, the EU-FAO Forest Law Enforcement, Governance and Trade Programme of the Food and Agriculture Organization of the United Nations published the study *The Voluntary Partnership Agreement (VPA) process in Central and West Africa: from theory to practice* to document and foster strategic reflection in partner countries already engaged in negotiating a VPA - or those who will be entering into such negotiations - by providing examples of good practices. These good practices were identified and recorded following interviews with the main stakeholders in the eight VPA countries in West and Central Africa, the European Forest Institute's (EFI) EU FLEGT Facility and the European Commission.

Saint Petersburg Declaration

The *Europe and North Asia Forest Law Enforcement and Governance* (ENA FLEG) Ministerial Conference was held in Saint Petersburg, Russia on 22–25 November 2005. In May 2004, the Russian Federation announced its intention to host the ENA FLEG process, supported by the World Bank. A preparatory conference was held in Moscow in June 2005.

The Saint Petersburg conference brought together nearly 300 participants representing 43 governments, the private sector, civil society and international organisations. It agreed to the *Saint Petersburg Declaration on Forest Law Enforcement and Governance in Europe and North Asia*. The Declaration includes an indicative list of actions, intended to serve as a general framework for possible actions to be undertaken by governments as well as civil society.

The conference took place as the United Kingdom prepared to pass the G8 Presidency to Russia. As Valery Roshchupkin, Head of the Federal Forestry Agency of the Russian Federation, confirmed, illegal logging would be of special importance for Russia as the G8 President and for the following G8 Summit, also held in Saint Petersburg.

United States

In response to growing concerns over illegal logging and advice from TRAFFIC and other organisations, on May 22, 2008 the U.S. amended the Lacey Act, when the Food, Conservation, and Energy Act of 2008 expanded its protection to a broader range of plants and plant products (Section 8204. Prevention of Illegal Logging Practices).

The requirements under the new Amendments are two-fold. First, the Lacey Act now makes it illegal to import into the United States plants that have been harvested contrary to any applicable Federal Law, State Law, Indian Tribal Law, or Foreign Law. If a plant is found to have been harvested in violation of the laws of the country where it was harvested, that plant would be subject to seizure and forfeiture if imported into the U.S. The Lacey Act also makes it unlawful, beginning December 15, 2008, to import certain plants and plant products without a Plant and Plant Product import declaration.

This Plant and Plant Product Declaration must contain (among other things) the Genus, Species, and Country of Harvest of every plant found in commercial shipments of certain products, a list of applicable products (along with other requirements and guidance) can be found on the USDA APHIS website.

Australia

The Timber Development Association (TDA) welcomes on June 6, 2014's release by the Australian Department of Agriculture of a position paper on the Illegal Logging Prohibition Regulation and guidance on how timber and wood products industry can comply on the Australian Government - Department of Agriculture official website. The release of the Government's guidance coincides with the release of industry developed timber due diligence tools and information through the industry website of Timber Due Diligence.

The Australian Illegal Logging Prohibition Regulation applies to importers into Australia of "regulated timber products" such as sawn timber, wood panels, pulp, paper products, and wood furniture. The Regulation starts on 30 November 2014 and requires that before import of these products or processing of raw logs, due diligence is undertaken to minimise the risk that the timber products or raw logs were illegally logged or incorporate illegally logged timber.

Deforestation and Climate Change

Deforestation is one of the main causes of climate change. It is the second largest anthropogenic source of carbon dioxide to the atmosphere, after fossil fuel combustion. Deforestation and forest degradation contribute to atmospheric greenhouse gas emissions through combustion of forest biomass and decomposition of remaining plant material and soil carbon. It used to account for more than 20% of carbon dioxide emissions, but it's currently somewhere around the 10% mark. By 2008, deforestation was 12% of total CO_2, or 15% if peatlands are included. These proportions are likely to have fallen since given the continued rise of fossil fuel use.

Averaged over all land and ocean surfaces, temperatures warmed roughly 1.53 °F (0.85 °C) between 1880 and 2012, according to the Intergovernmental Panel on Climate Change. In the Northern Hemisphere, 1983 to 2012 were the warmest 30-year period of the last 1400 years.

Effect on Climate Change

Decrease in Biodiversity

A 2007 study conducted by the National Science Foundation found that biodiversity and genetic diversity are codependent—that diversity among species requires diversity within a species, and vice versa. "If any one type is removed from the system, the cycle can break down, and the community becomes dominated by a single species."

Counteracting Climate Change

Reforestation

Reforestation is the natural or intentional restocking of existing forests and woodlands that have been depleted, usually through deforestation. It is the reestablishment of forest cover either naturally or artificially. Similar to the other methods of forestation, reforestation can be very effective because a single tree can absorb as much as 48 pounds of carbon dioxide per year and can sequester 1 ton of carbon dioxide by the time it reaches 40 years old.

Afforestation

Afforestation is the establishment of a forest or stand of trees in an area where there was no forest.

China

Although China has set official goals for reforestation, these goals were set for an 80-year time horizon and were not significantly met by 2008. China is trying to correct these problems with projects such as the Green Wall of China, which aims to replant forests and halt the expansion of the Gobi Desert. A law promulgated in 1981 requires that every school student over the age of 11 plant at least one tree per year. But average success rates, especially in state-sponsored plantings, remains relatively low. And even the properly planted trees have had great difficulty surviving the combined impacts of prolonged droughts, pest infestation and fires. Nonetheless, China currently has the highest afforestation rate of any country or region in the world, with 4.77 million hectares (47,000 square kilometers) of afforestation in 2008.

Japan

The primary goal of afforestation projects in Japan is to develop the forest structure of the nation and to maintain the biodiversity found in the Japanese wilderness. The Japanese temperate rainforest is scattered throughout the Japanese archipelago and is home to many endemic species that are not naturally found anywhere else. As development of the country's caused a decline in forest cover, a reduction in biodiversity was seen in those areas.

Agroforestry

Agroforestry or agro-sylviculture is a land use management system in which trees or shrubs are grown around or among crops or pastureland. It combines agricultural and forestry technologies to create more diverse, productive, profitable, healthy, and sustainable land-use systems.

Projects and Foundations

Arbor Day Foundation

Founded in 1972, the centennial of the first Arbor Day observance in the 19th century, the Foundation has grown to become the largest nonprofit membership organization dedicated to planting trees, with over one million members, supporters, and valued partners. They work on projects focused on planting trees around campuses, low-income communities, and communities that have been affected by natural disasters among other places.

Billion Tree Campaign

The Billion Tree Campaign was launched in 2006 by the United Nations Environment Programme (UNEP) as a response to the challenges of global warming, as well as to a wider array of sustainability challenges, from water supply to biodiversity loss. Its initial target was the planting of one billion trees in 2007. Only one year later in 2008, the campaign's objective was raised to 7 billion trees—a target to be met by the climate change conference that was held in Copenhagen, Denmark in December 2009. Three months before the conference, the 7 billion planted trees mark had been surpassed. In December 2011, after more than 12 billion trees had been planted, UNEP formally handed management of the program over to the not-for-profit Plant-for-the-Planet initiative, based in Munich, Germany.

The Amazon Fund (Brazil)

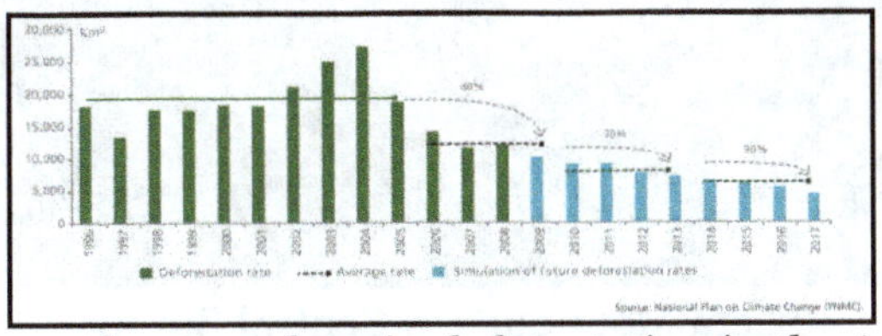

Four-year plan to reduce in deforestation in the Amazon.

Considered the largest reserve of biological diversity in the world, the Amazon Basin is also the largest Brazilian biome, taking up almost half the nation's territory. The Amazon Basin corresponds to two fifths of South America's territory. Its area of approximately seven million square kilometers covers the largest hydrographic network on the planet, through which runs about one fifth of the fresh water on the world's surface. Deforestation in the Amazon rainforest is a major cause to climate change due to the decreasing number of trees available to capture increasing carbon dioxide levels in the atmosphere.

The Amazon Fund is aimed at raising donations for non-reimbursable investments in efforts to prevent, monitor and combat deforestation, as well as to promote the preservation and sustainable use of forests in the Amazon Biome, under the terms of Decree N.º 6,527, dated August 1, 2008. The Amazon Fund supports the following areas: management of public forests and protected areas, environmental control, monitoring and inspection, sustainable forest management, economic activities created with sustainable use of forests, ecological and economic zoning, territorial arrangement and agricultural regulation, preservation and sustainable use of biodiversity, and recovery of deforested areas. Besides those, the Amazon Fund may use up to 20% of its donations to support the development of systems to monitor and control deforestation in other Brazilian biomes and in biomes of other tropical countries.

Forest Dieback

Forest dieback (also "Waldsterben", a German loan word) is a condition in trees or woody plants in which peripheral parts are killed, either by pathogens, parasites or due to conditions like acid rain and drought. Two of the nine tipping points for major climate changes, forecast for the next century, are directly related to forest diebacks. *Phomopsis azadirachtae* is a fungus of the Phomopsis genus that has been identified as the fungus responsible for dieback of *Azadirachta indica* (Neem) in India.

Jizera Mountains in Central Europe.

Definition

Forest dieback refers to the phenomena of a stand of trees losing health and dying without an obvious cause. This condition is also known as forest decline, forest damage, canopy level dieback, and stand level dieback (also Waldsterben and Waldschäden, German loan words). This usually affects individual species of trees, but can also affect multiple species. Trees and woody plants are killed by a complex group of factors that include parasites, such as fungus and beetles, pollution, such as acid rain and organic compounds, and drought. Diseases and pests can kill large groups of trees very easily, but the premature and progressive loss of clusters of trees without an obvious or specific cause is known as forest dieback. Dieback is an episodic event. Dieback takes on many locations and shapes. It can be along the perimeter, at specific elevations, or dispersed throughout the forest ecosystem.

Forest dieback has many symptoms: falling off of leaves and needles, discoloration of leaves and needles, thinning of the crowns of trees, dead stands of trees of a certain age, and changes in the roots of the trees. Forest dieback has many dynamic forms. A stand of trees can exhibit mild symptoms, extreme systems, or death at the same time. Forest decline can be viewed as the result of continued, widespread, and severe dieback of multiple species in the forest. Current forest decline is described by rapid development on individual trees, occurrence in different forest types, a long duration (over 10 years), and occurrence throughout the natural range of affected species.

History

Much research was done in the 1980s when a severe dieback happened in Germany and the Northeast United States. Previous diebacks were regionally limited, but starting at the end of the 1970s a decline took over the forests of Central Europe and parts of North

America. The forest damage in Germany was different because the decline was severe, up to 50% of the trees, prolonged, over 5 years, and widespread across tree species. Affected trees went from 8% in 1982 to 50% in 1984 and stayed at 50% through 1987. Many hypotheses have been proposed for this dieback.

Drought and temperature have been causal factors in many diebacks. This condition is common in semi-arid regions and the trees are already stressed by dehydration. There have been reported diebacks in the 50% mortality range across multiple species in Africa, Asia, Europe, North America, South and Central America, and Australia. Notable drought related diebacks have occurred globally from the 1970s through the 2000s. A few drought related diebacks have occurred before this. For instance, a 23% mortality rate was reported between 1945 and 1993 in Senegal Africa. Trees stressed by drought and temperature are more susceptible to parasites.

In North America there have been five notable hardwood diebacks this century. These have occurred with the maturation of the forest and each episode has lasted about eleven years. The most severe temperate forest dieback was on white birch and yellow trees. The others are ash, oak, and maple species. White and yellow birches experienced an episode starting between 1934 and 1957 and ending between 1953 and 1954. This followed a wave pattern that first appeared in southern regions and moved to northern regions. A second wave is evident between 1957 and 1965 in northern Quebec. Sugar maple experienced a wave of dieback in parts of the United States during the 1960s. A second wave occurred primarily in Canada during the 1980s, but also reached the United States. These diebacks were numerically analyzed to exclude natural tree mortality. It is hypothesized that a mature forest is more susceptible to extreme environmental stresses.

Pathogens are responsible for many diebacks. It is difficult to isolate and identify exactly which pathogens are responsible and how they interact with the trees. For instance *Phomopsis azadirachtae* Sateesh, Bhat, and Devaki are fungus of the *Phomopsis* genus that has been identified as the fungus responsible for the dieback in *Azadirachta indica* (Neem) in the regions of India. Neem is a very hardy and drought tolerant tree, has antifungal and antibacterial properties, but it still has many diseases. Different pathogens affect every part of the tree: the twigs, shoots, leaves, roots, and bark. Scientists have isolated both the fungus and the toxin. They have identified the pathogen in the tissue of all affected trees. Proper knowledge of toxin chemistry and its role in pathogenesis requires further investigations. Some experts consider dieback as a group of diseases with incompletely understood origins influenced by factors which predispose trees under stress to invasion.

Top dying disease is a disease of *Heritiera fomes*, the dominant species of mangrove tree growing in the Sundarbans. This disease has become more prevalent since about 1970 and may be linked to an increase in the heavy metal concentration of the sediment deposited in the Ganges Delta.

Hypothesis

The components of a forest ecosystem are complex and identifying specific cause-effect relationships between dieback and the environment is difficult. Etiology is the science of identifying the causes of death. Because there is no one single and clear cause of dieback there are multiple hypothesis for the causes and effects of dieback. The following hypotheses were agreed upon between the scientific exchanges of Germany and the United States in 1988.

Soil-Acidification/Aluminum-Toxicity: As soil becomes more acidic there is a release of aluminum that damages the tree's roots. Observed effects are: reduction of uptake and transport of some cations, reduction in root respiration, damage to fine feeder roots and root morphology, and reduction in elasticity of the cell walls. This is proposed by Professor Bernhard Ulrich in 1979.

Complex High-Elevation Disease: The combination of high ozone levels, acid deposition and nutrient deficiencies at high elevations kill trees. High ozone concentrations damage the leaves and needles of trees. Nutrients are leached from the foliage. The chain of events is magnified over time. This is proposed by a group of professors: Bernhard Prinz, Karl Rehfuess, Heinz Zöttle.

Red-Needle Disease of Spruce: The symptoms are needle drop and crown thinning. Needles turn a rust color and fall off. This is caused by foliar fungi that are secondary parasites attacking weakened trees. This is proposed by Professor Karl Rehfuess.

General Stress: The increased level of atmospheric pollutants hurts the root system and accumulates toxins in new leaves. Pollutants alter the growth, reduce the photoshythetic activity, and reduce the formation of secondary metabolites. It is believed that low concentrations levels are toxic. This is proposed by a group of professors led by Peter Schütt.

Excess Nitrogen Deposition: The increased level of nitrogen and ammonium, a fertilizer, could have the following possible effects: it could inhibit beneficial fungi, delay chemical reactions, disurbe normal balances between shoot growth and root growth, increase soil leaching. There is no experimental proof. This is proposed by Carl Olaf Tamm.

Organic Air Pollutants: This is similar to the general stress hypothesis, but focuses on organic compounds. The three compounds seriously discussed are ethylene, aniline, and dinitrophenol. Even at low levels these organic chemical compounds have caused: abnormal dropping of foliage, twisted foliage, and killing of seedlings. This is proposed by Fritz Führ.

Global Climate Change

Changes in mean annual temperature and drought are major contributing factors to forest dieback. As more carbon is released from dead trees, especially in the Amazon and Boreal forests, more greenhouse gases are released into the atmosphere. Increased levels of greenhouse gases increase the temperature of the atmosphere. The feedback loop is reinforced and the biological adaptations of the species determine its survival. Projections for dieback vary, but the threat of global climate change only stands to increase the rate of dieback.

Scientists do not know the exact tipping points of climate change and can only estimate the timescales. When a tipping point- the critical threshold- is reached, a small change in human activity can have long-term consequences on the environment. Two of the nine tipping points for major climate changes forecast for the next century are directly related to forest diebacks. Scientists are worried that forest dieback in the Amazon rain forest and the Boreal evergreen forest will trigger a tipping point in the next 50 years.

References

- Ron Nielsen, The Little Green Handbook: Seven Trends Shaping the Future of Our Planet, Picador, New York (2006) ISBN 978-0-312-42581-4
- John F. Mongillo; Linda Zierdt-Warshaw (2000). Linda Zierdt-Warshaw, ed. Encyclopedia of environmental science. University of Rochester Press. p. 104. ISBN 978-1-57356-147-1
- Stenstrup, Allen (2010). Forests. Greensboro, North Carolina: Morgan Reynolds Publishing. p. 89. ISBN 978-1-59935-116-2
- Rudel, T.K. 2005 "Tropical Forests: Regional Paths of Destruction and Regeneration in the Late 20th Century" Columbia University Press ISBN 0-231-13195-X
- Leakey, Richard and Roger Lewin, 1996, The Sixth Extinction : Patterns of Life and the Future of Humankind, Anchor, ISBN 0-385-46809-1
- Taylor, Leslie (2004). The Healing Power of Rainforest Herbs: A Guide to Understanding and Using Herbal Medicinals. Square One. ISBN 9780757001444

Chapter 5

Products Derived from Forest Resources

The resources obtained from forests are finite and have to be utilized judiciously. The integration of forest ecology and biology, forest measurements, forest policies and administration to manage the resources extracted from forest is called forest resource management. This chapter deals with forest products and forest resource management in an extensive manner for a better understanding of the subjects.

Woodworking

Artists can use woodworking to create delicate sculptures.

Woodworking is the activity or skill of making items from wood, and includes cabinet making (Cabinetry and Furniture), wood carving, joinery, and carpentry.

History

Along with stone, clay and animal parts, wood was one of the first materials worked by early humans. Microwear analysis of the Mousterian stone tools used by the Neanderthals show that many were used to work wood. The development of civilization was closely tied to the development of increasingly greater degrees of skill in working these materials.

Ancient Egyptian woodworking.

Among early finds of wooden tools are the worked sticks from Kalambo Falls, Clacton-on-Sea and Lehringen. The spears from Schöningen (Germany) provide some of the first examples of wooden hunting gear. Flint tools were used for carving. Since Neolithic times, carved wooden vessels are known, for example, from the Linear Pottery culture wells at Kückhofen and Eythra.

Woodworking shop in Germany in 1568, the worker in front is using a bow saw, the one in the background is planing.

Examples of Bronze Age wood-carving include tree trunks worked into coffins from northern Germany and Denmark and wooden folding-chairs. The site of Fellbach-Schmieden in Germany has provided fine examples of wooden animal statues from the Iron Age Wooden idols from the La Tène period are known from a sanctuary at the source of the Seine in France.

Ancient Egypt

There is significant evidence of advanced woodworking in Ancient Egypt. Woodworking is depicted in many extant ancient Egyptian drawings, and a considerable amount of ancient Egyptian furniture (such as stools, chairs, tables, beds, chests) has been preserved. Tombs represent a large collection of these artefacts and the inner coffins found in the tombs were also made of wood. The metal used by the Egyptians for woodworking tools was originally copper and eventually, after 2000 BC bronze as ironworking was unknown until much later.

Commonly used woodworking tools included axes, adzes, chisels, pull saws, and bow drills. Mortise and tenon joints are attested from the earliest Predynastic period. These joints were strengthened using pegs, dowels and leather or cord lashings. Animal glue came to be used only in the New Kingdom period. Ancient Egyptians invented the art of veneering and used varnishes for finishing, though the composition of these varnishes is unknown. Although different native acacias were used, as was the wood from the local sycamore and tamarisk trees, deforestation in the Nile valley resulted in the need for the importation of wood, notably cedar, but also Aleppo pine, boxwood and oak, starting from the Second Dynasty.

Ancient Rome

Woodworking was essential to the Romans. It provided, sometimes the only, material for buildings, transportation, tools, and household items. Wood also provided pipes, dye, waterproofing materials, and energy for heat. Although most examples of Roman woodworking have been lost, the literary record preserved much of the contemporary knowledge. Vitruvius dedicates an entire chapter of his De architectura to timber, preserving many details. Pliny, while not a botanist, dedicated six books of his Natural History to trees and woody plants which provides a wealth of information on trees and their uses.

Ancient China

Damascene woodworkers turning wood for mashrabia and hookass, 19th century.

The progenitors of Chinese woodworking are considered to be Lu Ban (魯班) and his wife Lady Yun, from the Spring and Autumn Period. Lu Ban is said to have introduced the plane, chalk-line, and other tools to China. His teachings were supposedly left behind in the book *Lu Ban Jing* (魯班經, "Manuscript of Lu Ban"). Despite this, it is believed that the text was written some 1500 years after his death. This book is filled largely with descriptions of dimensions for use in building various items such as flower pots, tables, altars, etc., and also contains extensive instructions concerning Feng Shui. It mentions almost nothing of the intricate glue-less and nail-less joinery for which Chinese furniture was so famous.

Micronesian of Tobi, Palau, making a paddle for his *wa* with an adze.

Modern Day

With the advances in modern technology and the demands of industry, woodwork as a field has changed. The development of Computer Numeric Controlled (CNC) Machines, for example, has made us able to mass-produce and reproduce products, faster, with less waste, and often more complex in design than ever before. CNC Routers can carve complicated and highly detailed shapes into flat stock, to create signs or art. Rechargeable power tools speed up creation of many projects and require much less body strength than in the past, when boring multiple holes, for example. Skilled fine woodworking, however, remains a craft pursued by many. There remains demand for hand crafted work such as furniture and arts, however with rate and cost of production, the cost for consumers is much higher.

Materials

Historically, woodworkers relied upon the woods native to their region, until transportation and trade innovations made more exotic woods available to the craftsman. Woods are typically sorted into three basic types: hardwoods typified by tight grain and derived from broadleaf trees, softwoods from coniferous trees, and man-made materials such as plywood and MDF.

Typically furniture such as tables and chairs is made using solid stock, and cabinet/fixture makers employ the use of plywood and other man made panel products.

Paperboard

Corrugated fiberboard made from paperboard.

Paperboard is a thick paper-based material. While there is no rigid differentiation between paper and paperboard, paperboard is generally thicker (usually over 0.25 mm, 0.010 in, or 10 points) than paper. According to ISO standards, paperboard is a paper with a grammage above 224 g/m2, but there are exceptions. Paperboard can be single- or multi-ply. Paperboard can be easily cut and formed, is lightweight, and because it is strong, is used in packaging. Another end-use would be graphic printing, such as book and magazine covers or postcards. Sometimes it is referred to as cardboard, which is a generic, lay term used to refer to any heavy paper pulp–based board. Paperboard is also used in fine arts for creating sculptures.

Sculpture made of massive paperboard by Herbert Wetterauer.

History

In 1817, the first paperboard carton was produced in England. Folding cartons first emerged around the 1860s and were shipped flat to save space, ready to be set up by customers when they were required. 1879 saw the development of mechanical die cutting and creasing of blanks. In 1911 the first kraft sulphate mill was built in Florida and in 1915 the gable top milk carton was patented and in 1935 the first dairy plant was observed using them. Ovenable paperboard was introduced in 1974.

Terminology and Classification

Terminology and classifications of paperboard are not always uniform. Differences occur depending on specific industry, locale, and personal choice. In general, the following are often used:

- Boxboard or cartonboard – paperboard for folding cartons and rigid set-up boxes:
    - Folding boxboard (FBB) – a bending grade capable of being scored and bending without fracture.
    - Chipboard – a recycled, low quality board:
        - White lined chipboard (WLC) – a white, often clay-coated chipboard.
    - Kraft board – a strong virgin fiber board often used for beverage carriers. Often clay- coated for printing.
    - Laminated board – a lamination of paperboards and other materials, for example liquid packaging board.
    - Solid Bleached Board (SBB) or Solid Bleached Sulphate (SBS) – clean white board used for foods etc. Sulphate refers to the kraft process.
    - Solid unbleached board (SUB) – board made from unbleached chemical pulp.
- Containerboard – a type of paperboard manufactured for the production of corrugated fiberboard:
    - Corrugated medium – the inner fluted portion of corrugated fiberboard.
    - Linerboard – a strong stiff board for one or both sides of corrugated boxes. It is the flat covering over the corrugating medium.
- Other:
    - Binder's board – a paperboard used in bookbinding for making hardcovers.

Production

Fibrous material is turned into pulp and bleached, to create one or more layers of board, which can be optionally coated for a better surface and/or improved visual appearance. Paperboard are produced on paper machines that can handle higher grammages and several plies.

Raw Materials

The above-mentioned fibrous material can either come from fresh (virgin) sources (e.g. wood) or from recycled waste paper. Around 90% of virgin paper is made from wood pulp. Today paperboard packaging in general, and especially products from certified sustainable sources, are receiving new attention, as manufacturers dealing with environmental, health, and regulatory issues look to renewable resources to meet increasing demand. It is now mandatory in many countries for paper-based packaging to be manufactured wholly or partially from recycled material.

Raw materials include:

- Hardwood: Ca. 0.05 inches (1.3 mm) in length e.g. Birch which has short fibres. It is generally more difficult to work with; however, it does provide higher tensile strength, but lower tear and other strength properties. Although its fibres are not as long and strong as those in softwood, they make for a stiffer product defined by some stifness tests. Hardwood fibres fill the sheet better and therefore make a smoother paper that is more opaque and better for printing. Hardwood makes an excellent corrugating medium.
- Softwood: Ca. 0.13 inches (3.3 mm) in length e.g. Pine and spruce which have typically long fibres and make superior paperboard in services where strength is important. Softwood makes excellent linerboard.
- Recycled: Used paper is collected and sorted and usually mixed with virgin fibres in order to make new material. This is necessary as the recycled fibre often loses strength when reused; the added virgin fibres enhance strength. Mixed waste paper is not usually deinked (skipping the deinking stage) for paperboard manufacture and hence the pulp may contain traces of inks, adhesives, and other residues which together give it a grey colour. Products made of recycled board usually have a less predictable composition and poorer functional properties than virgin fibre-based boards. Health risks have been associated with using recycled material in direct food contact. Swiss studies have shown that recycled material can contain significant portions of mineral oil, which may migrate into packed foods. Mineral oil levels of up to 19.4 mg/kg were found in rice packed in recycled board.
- Others: It is also possible to use the fibres of Sugarcane Bagasse, Straw, Hemp, Cotton, Flax, Kenaf, Abaca and other plant products.

Pulping

Two principal methods for extracting fibres from their sources are:

- Chemical pulping uses chemical solutions to convert wood into pulp, yielding around 30% less than mechanical pulping; however, pulp made by the kraft process has superior strength.
- Thermo mechanical pulp is a two-stage process which results in a very high yield of wood fibres at the expense of strength.

Bleaching

Pulp used in the manufacture of paperboard can be bleached to decrease colour and increase purity.

Virgin fibre pulp is naturally brown in colour, because of the presence of lignin. Recycled paperboard may contain traces of inks, bonding agents and other residue which colors it grey. Although bleaching is not necessary for all end-uses, it is vital for many graphical and packaging purposes. There are various methods of bleaching, which are used according to a number of factors for example, the degree of colour change required, chemicals chosen and method of treatment. There are three categories of bleaching methods:

- Bleaching by delignification using chlorine gas, which is a method that has been largely replaced by procedures which are gentler to the environment such as the use of oxygen as a replacement for the chlorine gas.
- Bleaching by oxidation using chemicals such as chlorine dioxide, hydrogen peroxide or sodium hypochlorite.
- Bleaching by reduction using chemicals such as sodium bisulphite.

Plies

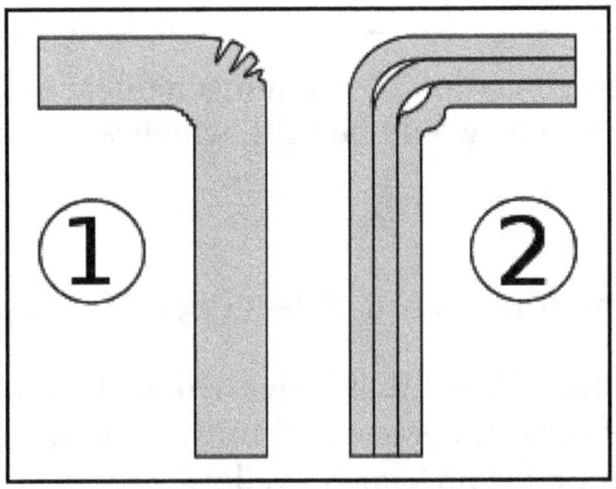

(1) single-ply vs. (2) multi-ply.

Multi-ply paperboard generally has higher creasing and folding performance than single-ply as a result of layering different types of pulp into a single product. In cases where the same kind of pulp is being used in several layers, each separate layer is treated and shaped individually in order to create the highest possible quality.

Coating

In order to improve whiteness, smoothness and gloss of paperboard, one or more layers of coating is applied. Coated paper is usually made up of:

- a pigment, which could be china clay, calcium carbonate or titanium dioxide, usually a combination of two.
- an adhesive or binder as styrene-butadiene emulsions or starches and water.

Additional components could be OBA (optical brightening agents).

Grades

The DIN Standard 19303 "Paperboard - Terms and grades" (Publication date : 2005-09) defines different grades of paperboard based on the surface treatment (first letter), the main furnish (second letter) and the colour (non-D grade) or bulk (D grade only) (numbering).

First letter (surface treatment)	Second letter (main furnish)	Number
• A = cast-coated • G = pigment coated • U = uncoated	• Z = bleached virgin chemical pulp • C = virgin mechanical pulp • N = unbleached virgin chemical pulp • T = recycled/secondary fibre with white, cream or brown reverse • D = recycled/secondary fibre with grey back	All except D grades: 1. white reverse side 2. cream reverse side 3. brown reverse side D grades only: 1. bulk $\geq 1.45\ cm^2/g$ 2. $1.3\ cm^2/g < bulk < 1.45\ cm^2/g$ 3. bulk $\leq 1.3\ cm^2/g$

Example: GC1 would be a "pigment coated", "virgin mechanical pulp" board with a "white reverse side". Often the used paperboard type would be FBB, which was coated on both sides.

Common Terms

Basis Weight (US): Is the weight in 1,000 square feet (93 m^2) of paperboard.

Brightness: Brightness is a technical term that is defined as the amount of blue-white light that a paper reflects. This property is very subjective and individual to each buyer and end use, as skin colour and food are better reproduced on 'warm' (yellow) whites and not blue whites.

Grammage: The grammage of the paperboard is assessed in accordance ISO 536. Grammage expresses mass per unit area and is measured in g/m^2.

PH: Surface pH is measured on a water extract and is on a scale of 0–14. 0 is acidic, 7 is neutral and 14 is alkaline.

Stiffness: Stiffness is one of the most important properties of paperboard as it affects the ability of cartons to run smoothly through the machine that erects, fills and closes them. Stiffness also gives strength and reduces the propensity of a carton to bulge under the weight of settling flowable contents such as cereals.

Although most paper strength properties increase with increasing sheet density, stiffness does not. A rule of thumb is that stiffness is proportional to the 1.6 power of sheet caliper.

The species of fiber used has an effect on stiffness, other things being equal. Northern softwood species impart superior stiffness compared to southern softwoods.

Other factors which affect board stiffness include coatings and moisture content.

Smoothness: Smoothness is particularly important when being used for printing, the smoother the paperboard, the better the image quality, because of better ink coverage. Smoothness is measured using air leak methods – the greater the rate of air leakage, at a specific air pressure, from under a cylindrical knife placed on the surface, the rougher the surface.

Caliper/Thickness: In the United States caliper is usually expressed in thousandths of an inch (0.001") or points, where a sheet of paperboard with a thickness of 0.024" would be 24 points. In Europe it is often sold in g/m^2, however the thickness of the board is measured in micron (µm).

Paperboard also tends to be referred to with thickness rather than weight.

Whiteness: It refers ideally to the equal presence of all colours, because a truly white sheet will reflect all wavelengths of visible light equally.

Paperboard Industry

The paperboard sector is mainly looked at in conjunction with the paper industry. The Paper & Paperboard market size (2007) had a value of 630.9 billion USD and a volume of 320.3 million metric tons . 40.1% of that market is European. About 50% of all produced paper is used for packaging, followed by printing and writing. According to ProCarton, the consumption of paper and paperboard seem to correlate with economic trends (GDP). Sales of carton in Europe sum up to around 8 billion Euros worth. Over 1,100 printers produce 5.4 million tonnes of cartonboard yearly. Cartons make up one third of paper and board packaging and 15% of all packaging. A bit more than half (54%) of the European carton is produced using recovered fibre or waste paper.

The paper and paperboard industry is quite energy and capital intensive. Just a coated board machine itself can cost around 90 - 120 million Euros (about 125 - 166 million USD in 11/2011). Economies of scale apply, because of which a few large players often dominate the market place. E.g. in North America the top 5 producers have a market share of 85%.

Fiberboard

Fiberboard (American spelling) or fibreboard (Commonwealth spelling) is a type of engineered wood product that is made out of wood fibers. Types of fiberboard (in order of increasing density) include particle board, medium-density fiberboard (MDF), and hardboard.

Fiberboard is sometimes used as a synonym for particle board, but particle board usually refers to low-density fiberboard. Plywood is not a type of fiberboard, as it is made of thin sheets of wood, not wood fibers or particles. Fiberboard, particularly medium-density fiberboard, is heavily used in the furniture industry. For pieces that will be visible, a veneer of wood is often glued onto fiberboard to give it the appearance of conventional wood.

In the packaging industry, the term "fiberboard" is often used to describe a tough kraft-based paperboard or corrugated fiberboard for boxes.

"Fiberboard" is also an intermediate product, an output of a pulp mill used as input for a paper mill.

Manufacture

Urea-formaldehyde (UF) resins are dominantly used in the medium density fiberboard (MDF) industry because of their low cost and fast curing characteristics. However, pressures on the use of UF resins are mounting steadily due to potential problems associated with formaldehyde emission. On the other hand, phenol-formaldehyde (PF) resins are more durable and do not emit formaldehyde after cure. The industry has traditionally shied away from using PF resins due primarily to their higher cost and much slower curing rate than UF resins. However, the press times for PF-bonded fiberboard can be substantially reduced by manipulating the fiber mat temperatures, molecular weight distribution of PF resins and pressing parameters. As a result, the press times for PF-bonded fiberboard can be made comparable to those for UF-bonded fiberboard. Also, the resin content required for PF-bonded fiberboard is less than 5% to achieve a good board quickly. This is considerably lower than that required for UF-bonded fiberboard.

Certain types of fiberboard can be considered "green" building products. Consisting of bio-based, secondary raw materials (wood chip or sugarcane fibers) recovered from within 100 miles (160 km) of manufacturing facilities, the binding agent used in this type of fiberboard is an all-natural product, consisting of vegetable starch containing no added formaldehydes.

Use

Fiberboard, classified by ASTM C208, Standard Specification for Cellulosic Fiber Insulating Board, has many benefits and is used in residential and commercial construction.

Applications include:

- Sound proofing/deadening,
- Structural sheathing,
- Low-slope roofing,
- Sound deadening flooring underlayment.

Fiberboard is also used in the automotive industry to create free-form shapes such as dashboards, rear parcel shelves, and inner door shells. These pieces are usually covered with a skin, foil, or fabric such as cloth, suede, leather, or polyvinyl chloride.

RSI Direct, A bi-weekly e-newsletter covering the roofing, siding and insulation industries, promotes the use of fiberboard as a coverboard in roof systems:

High density coated wood fiber is an ideal cover board, and the industry apparently agrees. More than two billion square feet of this product have already been installed in the U.S. roofing market. In terms of cost and availability, wood fiber is hard to beat.

Wood Veneer

In woodworking, veneer refers to thin slices of wood, usually thinner than 3 mm (1/8 inch), that typically are glued onto core panels (typically, wood, particle board or medium-density fiberboard) to produce flat panels such as doors, tops and panels for cabinets, parquet floors and parts of furniture. They are also used in marquetry. Plywood consists of three or more layers of veneer, each glued with its grain at right angles to adjacent layers for strength. Veneer beading is a thin layer of decorative edging placed around objects, such as jewelry boxes. Veneer is also used to replace decorative papers in Wood Veneer HPL. Veneer is also a type of manufactured board.

Wood veneer RCA Dimensia console TV set.

Veneer is obtained either by "peeling" the trunk of a tree or by slicing large rectangular blocks of wood known as flitches. The appearance of the grain and figure in wood comes from slicing through the growth rings of a tree and depends upon the angle at which the wood is sliced. There are three main types of veneer-making equipment used commercially:

- A rotary lathe in which the wood is turned against a very sharp blade and peeled off in one continuous or semi-continuous roll. Rotary-cut veneer is mainly used for plywood, as the appearance is not desirable because the veneer is cut concentric to the growth rings.

- A slicing machine in which the flitch or piece of log is raised and lowered against the blade and slices of the log are made. This yields veneer that looks like sawn pieces of wood, cut across the growth rings; such veneer is referred to as "crown cut".

- A half-round lathe in which the log or piece of log can be turned and moved in such a way as to expose the most interesting parts of the grain, creating a more textured feel and appearance; such veneer is commonly referred to as "rift cut."

Each slicing processes gives a very distinctive type of grain, depending upon the tree species. In any of the veneer-slicing methods, when the veneer is sliced, a distortion of the grain occurs. As it hits the wood, the knife blade creates a "loose" side where the cells have been opened up by the blade, and a "tight" side.

Historically veneers were also sawn, but this is more wasteful of wood. Veneering is an ancient art, dating back to at least the ancient Egyptians who used expensive and rare wood veneers over cheaper timbers to produce their furniture and sarcophagi. During the Roman Empire, Romans also used veneered work in mass quantities.

Producing Wood Veneers

The finest and rarest logs are sent to companies that produce veneer. The advantage to this practice is twofold. First, it provides the most financial gain to the owner of the log. Secondly, and of more importance to the woodworker, it greatly expands the amount of usable wood. While a log used for solid lumber is cut into thick pieces, usually no thinner than 1 1/8 inches (3 cm), veneers are cut as thin as 1/40 of an inch (0.6mm). Depending on the cutting process used by the veneer manufacturer, very little wood is wasted by the saw blade thickness, known as the saw kerf. Accordingly, the yield of a rare grain pattern or wood type is greatly increased, in turn placing less stress on the resource. Some manufacturers even use a very wide knife to "slice off" the thin veneer pieces. In this way, none of the wood is wasted. The slices of veneer are always kept in the order in which they are cut from the log and are often sold this way.

Types of Veneers

There are a few types of veneers available, each serving a particular purpose.

- Raw veneer has no backing on it and can be used with either side facing up. It is important to note that the two sides will appear different when a finish has been applied, due to the cell structure of the wood.
- Paper backed veneer is as the name suggests, veneers that are backed with paper. The advantage to this is it is available in large sizes, or sheets, as smaller pieces are joined together prior to adding the backing. This is helpful for users that do not wish to join smaller pieces of raw veneers together. This is also helpful when veneering curves and columns as the veneer is less likely to crack.
- Phenolic backed veneer is less common and is used for composite, or manmade wood veneers. Due to concern for the natural resource, this is becoming more popular. It too has the advantage of being available in sheets, and is also less likely to crack when being used on curves.
- Laid up veneer is raw veneer that has been joined together to make larger pieces. The process is time-consuming and requires great care, but is not difficult and requires no expensive tools or machinery. Veneers can be ordered through some companies already laid up to any size, shape or design.
- Reconstituted veneer is made from fast-growing tropical species. Raw veneer is cut from a log, and dyed if necessary. Once dyed, the sheets are laminated together to form a block. The block is then sliced so that the edges of the laminated veneer become the "grain" of the reconstituted veneer.
- Wood on Wood Also called 2-ply is a decorative wood veneer face with a utility grade wood backer applied at an opposing direction to the face veneer.

Advantages of Using Veneers

Compared to wood, one of the primary advantages of using veneer is stability. While solid wood can be prone to warping and splitting, because veneer is made of thin layers of wood glued together, the chances of splitting or cracking are reduced. Further, the glue used provides additional strength, making the end result stronger than natural wood.

Some projects built using wood veneer would not be possible to construct using solid lumber, owing to expansion and contraction caused by fluctuation of temperature and humidity. Another advantage of veneer is sustainability—furniture made with wood veneer uses less wood than the same piece of furniture made with solid wood. Further, veneer may also be more readily available than solid wood as exotic hardwood lumber can be scarce and very expensive.

Ecological Characteristics

- Recyclability and renewability: Wood has the least impact on total energy use, greenhouse gasses, air and water pollution, solid waste and ecological resource use. Seventy percent less energy is required in the use of wood compared with any other building materials.
- Sustainability: Using veneer extends the use of a piece of timber. The wood that might be used in one solid piece a few visible cm wide can cover a far greater area when used as a veneer.
- Toxicity: Nontoxic; veneer stores carbon and also maximises the use of harvested wood.

Buying Veneers

Wood veneers are typically sold by the square foot. With the ability to join veneers, even small pieces are usable, resulting in very little waste. Many sources sell small packets of veneers that are sequence matched and are ideal for small projects. These make experimenting and practicing much more economical. It is also possible to buy plywood and other substrates with veneered faces for larger projects consisting of casework.

Wood Preservation

All measures that are taken to ensure a long life of wood fall under the definition wood preservation (timber treatment).

Detail of sample in photo above.

In moist and oxygenated soil, there are few treatments that enable vulnerable wood (softwood here) to resist for long against bacterial or fungal degradation.

Apart from structural wood preservation measures, there are a number of different (chemical) preservatives and processes (also known as timber treatment, lumber treatment or pressure treatment) that can extend the life of wood, timber, wood structures or engineered wood. These generally increase the durability and resistance from being destroyed by insects or fungus.

History

As proposed by Richardson, treatment of wood has been practised for almost as long as the use of wood itself. There are records of wood preservation reaching back to ancient Greece during Alexander the Great's rule, where bridge wood was soaked in olive oil. The Romans protected their ship hulls by brushing the wood with tar. During the Industrial Revolution wood preservation became a cornerstone of the wood processing industry. Inventors and scientists such as Bethell, Boucherie, Burnett and Kyan made historic developments in wood preservation, with the preservative solutions and processes. Commercial pressure treatment began in the latter half of the 19th century with the protection of railroad crossties using creosote. Treated wood was used primarily for industrial, agricultural, and utility applications, where it is still used, until its use grew considerably (at least in the United States) in the 1970s as homeowners began building decks and backyard projects. Innovation in treated timber products continues to this day, with consumers becoming more interested in less toxic materials.

A modern wharf piling bored by bivalves known as *shipworms*.

Hazards

Wood that has been industrially pressure-treated with approved preservative products poses a limited risk to the public, and should be disposed of properly. On December 31, 2003, the U.S. wood treatment industry stopped treating residential lumber with arsenic and chromium (chromated copper arsenate, or CCA). This was a voluntary agreement with the United States Environmental Protection Agency. CCA was replaced by copper-based pesticides, with exceptions for certain industrial uses. CCA may still be used for outdoor products like utility trailer beds and non-residential construction like piers, docks, and agricultural buildings. Industrial wood preservation chemicals are generally not available directly to the public and may require special approval to import or purchase depending on the product and the jurisdiction where being used. In most countries, industrial wood preservation operations are notifiable industrial activities that require licensing from relevant regulatory authorities such as EPA or equivalent. Reporting and licensing conditions vary widely depending on the particular chemicals used and the country of use.

Although pesticides are used to treat lumber, preserving lumber protects natural resources (in the short term) by enabling wood products to last longer. Previous poor practices in industry have left legacies of contaminated ground and water around wood treatment sites in some cases. However, under currently approved industry practices and regulatory controls such as implemented in Europe, North America, Australia, New Zealand, Japan and elsewhere, environmental impact of these operations should be minimal.

Wood treated with modern preservatives is generally safe to handle given appropriate handling precautions and personal protection measures. However, treated wood may present certain hazards in some circumstances such as during combustion or where loose wood dust particles or other fine toxic residues are generated or where treated wood comes into direct contact with food and agriculture.

Preservatives containing copper in the form of microscopic particles have recently been introduced to the market, usually with "micronized" or "micro" trade names and designations such as MCQ or MCA. While the manufacturers represent that these products are safe and EPA has registered these products, some groups have expressed concerns regarding exposure to engineered sub-micron and nano-sized copper particles. These concerns have led to a debate among advocacy groups and governments on whether special regulation of nanotechnology is warranted. (See nanotechnology and nanotoxicology for additional information).

The American Wood Protection Association (AWPA), recommends that all treated wood be accompanied by a Consumer Information Sheet (CIS), to communicate safe handling and disposal instructions as well as potential health and environmental hazards of treated wood. Many producers have opted instead to provide Material Safety Data Sheets (MSDS) instead. Although the practice of distributing MSDS instead of CIS is widespread there is an ongoing debate regarding the practice and how best to best communicate potential hazards and hazard mitigation to the end-user. Neither MSDS or the newly adopted International Safety Data Sheets (SDS) are required for treated lumber under current U.S. Federal law.

Businesses and Organizations

American Wood Protection Association

Founded in 1904, the American Wood Protection Association (AWPA), formerly American Wood-Preservers' Association, is a non-profit organization which is the standard setting body for wood preservation standards (including ANSI). AWPA Standards are developed by its technical committees in an open, consensus-based process that involves individuals from all facets of wood preservation: Producers of preservatives and preservative components; producers of treated and untreated wood products; end users of treated wood; engineers, architects and building code officials; government entities, academia, and other groups with a general interest in wood preservation. AWPA's Standards are universally specified for wood preservation in the US, and are recognized worldwide.

AWPA standards help ensure that treated wood products perform satisfactorily for their intended use. They are recognized and used by most, if not all, specifiers of treated wood including electrical utility, marine, road and building construction as well as by local, state and federal governments. "AWPA", "American Wood Protection Association", identifiers of AWPA Standards (e.g., U1, T1, M4, etc.), and Use Category designations (e.g., UC1, UC3B, UC4A, etc.) are AWPA trademarks and the intellectual property of AWPA and its Technical Committees.

Wood preservative systems produced under the AWPA standards system for the residential market are required to be inspected under the stringent American Lumber Standards Committee (ALSC) third party inspection system in order to assure compliance with AWPA standards.

While many wood preservative systems are produced under the AWPA standards system, there are wood preservative products in the market that have not earned AWPA standard status and are not subject to the ALSC inspection system. Compliance with AWPA and ASLC will be noted by the AWPA logo on the product end tags.

Marketing of Wood Preservatives

In general, marketers of wood preservation systems favor certain terminology. For example, the term 'preservative' is used in preference to words such as: chemical, pesticide, fungicide or biocide. And with newer preservatives, the term 'micronized' is favored over the term nanoparticle or nanotechnology, which may raise public safety concerns.

Chemical Preservatives

Chemical preservatives can be classified into three broad categories: water-borne preservatives, oil-borne preservatives, and light organic solvent preservatives (LOSPs). These are discussed in more detail below.

Micronized Copper Technology

Particulate (micronised or dispersed) copper preservative technology has recently been introduced in the US and Europe. In these systems, the copper is ground to micro sized particles and suspended in water rather than being dissolved in a chemical reaction as is the case with other copper products such as ACQ and Copper Azole. There are currently two particulate copper systems in production. One system uses a quat biocide system (known as MCQ) and is a take-off of ACQ. The other uses an azole biocide (known as MCA or μCA-C) and is a take-off of Copper Azole.

Proponents of the particulate copper systems make the case that the particulate copper system perform as well or better than the dissolved copper systems as a wood preservative, but other industry researchers disagree. None of the particulate copper systems have been submitted to the American Wood Protection Association (AWPA) for evaluation, thus the particulate systems should not be used in applications where AWPA standards are required. However, all of the particulate copper systems have been tested and approved for building code requirements by the International Code Council (ICC). The particulate copper systems provide a lighter color than dissolved copper systems such as ACQ or copper azole.

Proponents of the micronized copper systems claim that the systems are subject to third party inspection under a quality monitor program. However, the monitoring program is not subject to oversight by the American Lumber Standards Committee (ALSC) as is required for the AWPA standard systems.

Two particulate copper systems, one marketed as MicroPro and the other as Wolmanized using μCA-C formulation, have achieved Environmentally Preferable Product (EPP) certification. The *EPP* certification was issued by Scientific Certifications Systems (SCS), and is based on a comparative life-cycle impact assessments with an industry standard.

The copper particle size used in the "micronized" copper beads ranges from 1 to 700 nm with an average under 300 nm. Larger particles (such as actual micron-scale particles) of copper do not adequately penetrate the wood cell walls. These micronized preservatives use nano particles of copper oxide or copper carbonate, for which there are alleged safety concerns. An environmental group has recently petitioned EPA to revoke the registration of the micronized copper products citing safety issues.

Alkaline Copper Quaternary

Alkaline copper quaternary (ACQ) is a preservative made of copper, a fungicide, and a quaternary ammonium compound (quat) like didecyl dimethyl ammonium chloride, an insecticide which also augments the fungicidal treatment. ACQ has come into wide use in the US, Europe, Japan and Australia following restrictions on CCA. Its use is governed by national and international standards, which determine the volume of preservative uptake required for a specific timber end use.

Since it contains high levels of copper, ACQ-treated timber is five times more corrosive to common steel. It is necessary to use double-galvanized or stainless steel fasteners in ACQ timber. Use of fasteners meeting or exceeding requirements for ASTM A 153 Class D meet the added requirements for fastener durability. The U.S. began mandating the use of non-arsenic containing wood preservatives for virtually all residential use timber in 2004.

The American Wood Protection Association (AWPA) standards for ACQ require a retention of 0.15 lb/ft^3 (PCF) for above ground use and 0.40 lb/ft^3 for ground contact.

Chemical Specialties, Inc (CSI, now Viance) received U.S. Environmental Protection Agency's Presidential Green Chemistry Challenge Award in 2002 for commercial introduction of ACQ. Its widespread use has eliminated major quantities of arsenic and chromium previously contained in CCA.

Copper Azole

Copper azole preservative (denoted as CA-B and CA-C under American Wood Protection Association/AWPA standards) is a major copper based wood preservative that has come into wide use in Canada, the US, Europe, Japan and Australia following restrictions on CCA. Its use is governed by national and international standards, which determine the volume of preservative uptake required for a specific timber end use.

Copper azole is similar to ACQ with the difference being that the dissolved copper preservative is augmented by an azole co-biocide like Tebuconazole instead of the quat biocide used in ACQ. The azole co-biocide yields a copper azole product that is effective at lower retentions than required for equivalent ACQ performance.

It is marketed widely under the *Wolmanized* brand in North America, and the *Tanalith* brand across Europe and other international markets.

The AWPA standard retention for CA-B is 0.10 lb/ft^3 for above ground applications and 0.21 lb/ft^3 for ground contact applications. Type C copper azole, denoted as CA-C, has been introduced under the Wolmanized brand. The AWPA standard retention for CA-C is 0.06 lb/ft^3 for above ground applications and 0.15 lb/ft^3 for ground contact applications.

The copper azole preservative incorporates organic triazoles such as tebuconazole or propiconazole as the co-biocide, which are also used to protect food crops. The general appearance of wood treated with copper azole preservative is similar to CCA with a green colouration.

Copper Naphthenate

Copper Naphthenate a Danish invention of 1911, has been used as a preservative and water repellant for more than 100 years. Copper Naphthenate has been used effectively for many applications including: fence posts, canvas, nets, greenhouses, utility poles, railroad ties, beehives, and wooden structures in ground contact. Copper Naphthenate is registered with the EPA as a non-restricted use pesticide, so there is no federal applicators licensing requirements for its use as a wood preservative. Copper Naphthenate can be applied by brush, dip, or pressure treatment.

The University of Hawaii has found that Copper Naphthenate in wood at loadings of 1.5 lbs per cubic foot is resistant to Formosan termite attack. On February 19, 1981 the Federal Register outlined the EPA's position regarding the health risks associated the various wood preservatives. As a result, the National Park Service recommended the use of copper naphthenate in its facilities as an approved substitute for pentachlorophenal, creosote, and inorganic arsenicals. A 50-year study presented to AWPA in 2005 by Mike Freeman and Douglas Crawford says, "This study reassessed the condition of the treated wood posts in southern Mississippi, and statistically calculated the new expected post life span. It was determined that commercial wood preservatives, like pentachlorophenol in oil, creosote, and copper naphthenate in oil, provided excellent protection for posts, with life spans now calculated to exceed 60 years. Surprisingly, creosote and penta treated posts at 75% of the recommended AWPA retention, and Copper Naphthenate at 50% of the required AWPA retention, gave excellent performance in this AWPA Hazard Zone 5 site. Untreated southern pine posts lasted 2 years in this test site."

The AWPA M4 Standard for the care of preservative-treated wood products, reads, "The appropriateness of the preservation system for field treatment shall be determined by the type of preservative originally used to protect the product and the availability of a field treatment preservative. Because many preservative products are not packaged and labeled for use by the general public, a system different from the original treatment may need to be utilized for field treatment. Users shall carefully read and follow the instructions and precautions listed on the product label when using these materials. Copper Naphthenate preservatives containing a minimum of 2.0% copper metal are recommended for material originally treated with copper naphthenate, pentachlorophenol, creosote, creosote solution or waterborne preservatives." The M4 Standard has been adopted by the International Code Council's (ICC) International Building Code (IBC) section 2303.1.8 Preservative-treated Wood, and International Residential Code (IRC) R319.1.1 Field Treatment. The American Association of State Highway and Transportation Officials AASHTO has also adopted the AWPA M4 Standard.

A waterborne copper naphthenate is sold to consumers under the tradename QNAP 5W. Oilborne copper napthenates with 1% copper as metal solutions are sold to consumers under the tradenames Copper Green, and Wolmanized Copper Coat, a 2% copper as metal solution is sold under the tradename Tenino.

Chromated Copper Arsenate (CCA)

In CCA treatment, copper is the primary fungicide, arsenic is a secondary fungicide and an insecticide, and chromium is a fixative which also provides ultraviolet (UV) light resistance. Recognized for the greenish tint it imparts to timber, CCA is a preservative that was extremely common for many decades.

In the pressure treatment process, an aqueous solution of CCA is applied using a vacuum and pressure cycle, and the treated wood is then stacked to dry. During the process, the mixture of oxides reacts to form insoluble compounds, helping with leaching problems.

The process can apply varying amounts of preservative at varying levels of pressure to protect the wood against increasing levels of attack. Increasing protection can be applied (in increasing order of attack and treatment) for: exposure to the atmosphere, implantation within soil, or insertion into a marine environment.

In the last decade concerns were raised that the chemicals may leach from the wood into surrounding soil, resulting in concentrations higher than naturally occurring background levels. A study cited in *Forest Products Journal* found 12–13% of the chromated copper arsenate leached from treated wood buried in compost during a 12-month period. Once these chemicals have leached from the wood, they are likely to bind to soil particles, especially in soils with clay or soils that are more alkaline than neutral. In the United States the US Consumer Product Safety Commission issued a report in 2002 stating that exposure to arsenic from direct human contact with CCA treated wood may be higher than was previously thought. On 1 January 2004, the Environmental Protection Agency (EPA) in a voluntary agreement with industry began restricting the use of CCA in treated timber in residential and commercial construction, with the exception of shakes and shingles, permanent wood foundations, and certain commercial applications. This was in an effort to reduce the use of arsenic and improve environmental safety, although the EPA were careful to point out that they had not concluded that CCA treated wood structures in service posed an unacceptable risk to the community. The EPA did not call for the removal or dismantling of existing CCA treated wood structures.

In Australia, the Australian Pesticides and Veterinary Medicines Authority (APVMA) restricted the use of CCA preservative for treatment of timber used in certain applications from March 2006. CCA may no longer be used to treat wood used in 'intimate human contact' applications such as children's play equipment, furniture, residential decking and handrailing. Use for low contact residential, commercial and industrial applications remains unrestricted, as does its use in all other situations. The APVMA decision to restrict the use of CCA in Australia was a precautionary measure, even though the report found no evidence that demonstrated CCA treated timber posed unreasonable risks to humans in normal use. Similarly to the US EPA, the APVMA did not recommend dismantling or removal of existing CCA treated wood structures.

In Europe, Directive 2003/2/EC restricts the marketing and use of arsenic, including CCA wood treatment. CCA treated wood is not permitted to be used in residential or domestic constructions. It is permitted for use in various industrial and public works, such as bridges, highway safety fencing, electric power transmission and telecommunications poles. In the United Kingdom waste timber treated with CCA was classified in July 2012 as hazardous waste by the Department for the Environment, Food and Rural Affairs.

Other Copper Compounds

These include copper HDO (Bis-(N-cyclohexyldiazeniumdioxy)-copper or CuHDO), copper chromate, copper citrate, acid copper chromate, and ammoniacal copper zinc arsenate (ACZA). The CuHDO treatment is an alternative to CCA, ACQ and CA used in Europe and in approval stages for United States and Canada. ACZA is generally used for marine applications.

Borate Preservatives

Boric acid, oxides and salts (borates) are effective wood preservatives and are supplied under numerous brand names throughout the world. One of the most common compounds used is disodium octaborate tetrahydrate (commonly abbreviated DOT). Borate treated wood is of low toxicity to humans, and does not contain copper or other heavy metals. However, unlike most other preservatives, borate compounds do not become fixed in the wood and can be partially leached out if exposed repeatedly to water that flows away rather than evaporating (evaporation leaves the borate behind so is not a problem). Even though leaching will not normally reduce boron concentrations below effective levels for preventing fungal growth, borates should not be used where they will be exposed to repeated rain, water or ground contact unless the exposed surfaces are treated to repel water. Zinc-borate compounds are less suspectible to leaching than sodium-borate compounds, but are still not recommended for below-ground use unless the timber is first sealed. Recent interest in low toxicity timber for residential use, along with new regulations restricting some wood preservation agents, has resulted in a resurgence of the use of borate treated wood for floor beams and internal structural members. Researchers at CSIRO in Australia have developed organoborates which are much more resistant to leaching, while still providing timber with good protection from termite and fungal attack. The cost of the production of these modified borates will limit their widespread take-up but they are likely to be suitable for certain niche applications, especially where low mammalian toxicity is of paramount importance.

PTI Wood Preservatives

Recent concerns about the health and environmental impacts of metallic wood preservatives have created a market interest in non-metallic wood preservatives such as Propiconazole-Tebuconazole-Imidicloprid better known as PTI. The American Wood Protection Association (AWPA) standards for PTI require a retention of 0.018 lb/ft3 (PCF) for above ground use and 0.013 lb/ft3 when applied in combination with a wax stabilizer. The AWPA has not developed a standard for a PTI ground contact preservative, so PTI is currently limited to above ground applications such as decks. All three of the PTI components are also used in food crop applications. The very low required retentions for PTI pressure treated wood further limits impacts plus substantinally decreases the freight costs and associated environmental impacts for shipping preservative components to the pressure treating plants.

The PTI preservative imparts very little color to the wood. Producers generally add a color agent or a trace amount of copper solution so as to identify the wood as pressure treated and to better match the color of other pressure treated wood products. The PTI wood products are very well adapted for paint and stain applications with no bleed-through. The addition of the wax stabilizer allows a lower preservative retention plus substantially reduces the tendency of wood to warp and split as it dries. In combination with normal deck maintenance and sealer applications, the stabilizer helps maintain appearance and performance over time. PTI pressure treated wood products are no more corrosive than untreated wood and are approved for all types of metal contact, including aluminum.

PTI pressure treated wood products are relatively new to the market place and are not yet widely available in building supply stores. However, there are some suppliers selling PTI products for delivery anywhere in the US on a job lot order basis.

Sodium Silicate-based Preservatives

Sodium silicate is produced by fusing sodium carbonate with sand or heating both ingredients under pressure. It has been in use since the 19th century. It can be a deterrent against insect attack and possesses minor flame-resistant properties; however, it is easily washed out of wood by moisture, forming a flake-like layer on top of the wood.

Timber Treatment Technology, LLC, markets TimberSIL®, a sodium silicate wood preservative. The TimberSIL® proprietary process surrounds the wood fibers with a protective, non-toxic, amorphous glass matrix. The result is a product the company calls “Glass Wood,” a material that is Class A fire-retardant, chemically inert, rot and decay resistant, and superior in strength to untreated wood.Another option is to use plywood, fire resistant plywood.

Potassium Silicate-based Preservatives

There are a number of European natural paint fabricants that have developed potassium silicate (potassium waterglass) based preservatives. They frequently include boron compounds, cellulose, lignin and other plant extracts. They are a surface application with a minimal impregnation for internal use.

Bifenthrin Spray Preservatives

In Australia, a water-based bifenthrin preservative has been developed to improve the insect resistance of timber. As this preservative is applied by spray, it only penetrates the outer 2 mm of the timber cross-section. Concerns have been raised as to whether this thin-envelope system will provide protection against insects in the longer term, particularly when exposed to sunlight for extended periods.

Fire Retardant Treated

This treated wood utilizes a fire retardant chemical that remains stable in high temperature environments. The fire retardant is applied under pressure at a wood treating plant like the preservatives described above, or applied as a surface coating.

In both cases, treatment provides a physical barrier to flame spread. The treated wood chars but does not oxidize. Effectively this creates a convective layer that transfers flame heat to the wood in a uniform way which significantly slows the progress of fire to the material. There are several commercially available wood-based construction materials using pressure-treatment (such as those marketed in the United States and elsewhere under the trade names of "FirePro", "Burnblock" 'Woodsafe, Dricon', 'D-Blaze,' and 'Pyro-Guard'), as well as factory-applied coatings under the trade names of 'PinkWood' and 'BluWood'. Some site-applied coatings as well as brominated fire retardants have lost favor due to safety concerns as well as concerns surrounding the consistency of application. Specialized treatments also exist for wood used in weather-exposed applications.

The only impregnation-applied fire retardant commercially available in Australia (marketed under the tradename 'Guardian') used calcium formate as a 'powerful wood modifying agent', but was removed from sale in early 2010 for unspecified reasons.

Oil-borne Preservatives

These include pentachlorophenol ("penta") and creosote. They emit a strong petrochemical odor and are generally not used in consumer products.

Coal-tar Creosote

Creosote was the first wood preservative to gain industrial importance more than 150 years ago and it is still widely used today for protection of industrial timber components where long service life is essential. Creosote is a tar-based preservative that is commonly used for utility poles and railroad ties (UK: railway sleepers). Creosote is one of the oldest wood preservatives, and was originally derived from a wood distillate, but now, virtually all creosote is manufactured from the distillation of coal tar. Creosote is regulated as a pesticide, and is not usually sold to the general public.

Linseed Oil

In recent years in Australia and New Zealand, linseed oil has been incorporated in preservative formulations as a solvent and water repellent to "envelope treat" timber. This involves just treating the outer 5 mm of the cross-section of a timber member with preservative (e.g., permethrin 25:75), leaving the core untreated. While not as effective as CCA or LOSP methods, envelope treatments are significantly cheaper, as they use far less preservative. Major preservative manufacturers add a blue (or red) dye to envelope treatments. Blue colored timber is for use south of the Tropic of Capricorn and red for elsewhere. The colored dye also indicates that the timber is treated for resistance to termites/white ants. There is an ongoing promotional campaign in Australia for this type of treatment.

Other Emulsions

Light Organic Solvent Preservatives (LOSP)

This class of timber treatments use white spirit, or light oils such as kerosene, as the solvent carrier to deliver preservative compounds into timber. Synthetic pyrethroids are typically used as an insecticide, such as permethrin, bifenthrin or deltamethrin. In Australia and New Zealand, the most common formulations use Permethrin as an insecticide, and Propaconazole and Tebuconazole as fungicides. While still using a chemical preservative, this formulation contains no heavy-metal compounds.

With the introduction of strict volatile organic compound (VOC) laws in the European Union, LOSPs have disadvantages due to the high cost and long process times associated with vapour-recovery systems. LOSPs have been emulsified into water-based solvents. While this does significantly reduce VOC emissions, the timber swells during treatment, removing many of the advantages of LOSP formulations.

Epoxy

Various Epoxy resins usually thinned with a solvent like acetone or MEK can be used to both preserve and seal wood.

New Technologies

Wood Acetylation

This bridge made from acetylated wood near Sneek, the Netherlands, is designed to carry heavy traffic.

Chemical modification of wood at the molecular level has been used to improve its performance properties. Many chemical reaction systems for the modification of wood, especially those using various types of anhydrides, have been published; however, the reaction of wood with acetic anhydride has been the most studied.

The physical properties of any material are determined by its chemical structure. Wood contains an abundance of chemical groups called *free hydroxyls*. Free hydroxyl groups readily absorb and release water according to changes in the climatic conditions to which they are exposed. This is the main reason why wood's dimensional stability is impacted by swelling and shrinking. It is also believed that the digestion of wood by enzymes initiates at the free hydroxyl sites, which is one of the principal reasons why wood is prone to decay.

Acetylation effectively changes the free hydroxyls within wood into acetyl groups. This is done by reacting the wood with acetic anhydride, which comes from acetic acid. When free hydroxyl groups are transformed to acetyl groups, the ability of the wood to absorb water is greatly reduced, rendering the wood more dimensionally stable and, because it is no longer digestible, extremely durable. In general, softwoods naturally have an acetyl content from 0.5 to 1.5% and more durable hardwoods from 2 to 4.5%. Acetylation takes wood well beyond these levels with corresponding benefits. These include an extended coatings life due to acetylated wood acting as a more stable substrate for paints and translucent coatings. Acetylated wood is non-toxic and does not have the environmental issues associated with traditional preservation techniques.

The acetylation of wood was first done in Germany in 1928 by Fuchs. In 1946, Tarkow, Stamm and Erickson first described the use of wood acetylation to stabilize wood from swelling in water. Since the 1940s, many laboratories around the world have looked at acetylation of many different types of woods and agricultural resources.

In spite of the vast amount of research on chemical modification of wood, and, more specifically, on the acetylation of wood, commercialization did not come easily. The first patent on the acetylation of wood was filed by Suida in Austria in 1930. Later, in 1947, Stamm and Tarkow filed a patent on the acetylation of wood and boards using pyridine as a catalyst. In 1961, the Koppers Company published a technical bulletin on the acetylation of wood using no catalysis, but with an organic cosolvent In 1977, in Russia, Otlesnov and Nikitina came close to commercialization, but the process was discontinued, presumably because cost-effectiveness could not be achieved. In 2007, Titan Wood, a London-based company, with production facilities in The Netherlands, achieved cost-effective commercialization and began large-scale production of acetylated wood under the trade name "Accoya".

Natural Preservatives

Copper Plating

Copper plating or Copper sheathing is the practice of covering wood most usually wooden hulls of ships with copper metal. As metallic copper is both repellent and toxic to fungus, insects such as termites, and marine bi-valves this would preserve the wood and also act as an anti-fouling measure to prevent aquatic life from attaching to the ship's hull and reducing a ship's speed and maneuverability.

Naturally Rot-resistant Woods

These species are resistant to decay in their natural state, due to high levels of organic chemicals called *extractives*, mainly polyphenols. Extractives are chemicals that are deposited in the heartwood of certain tree species as they convert sapwood to heartwood. Huon pine (*Lagarostrobos franklinii*), merbau (*Intsia bijuga*), ironbark (*Eucalyptus* spp.), tōtara (*Podocarpus totara*), puriri (*Vitex lucens*), kauri (*Agathis australis*), and many cypresses, such as coast redwood (*Sequoia sempervirens*) and western red cedar (*Thuja plicata*), fall in this category. However, many of these species tend to be prohibitively expensive for general construction applications.

Huon pine was used for ship hulls in the 19th century, but over-harvesting and Huon pine's extremely slow growth rate makes this now a speciality timber. Huon pine is so rot resistant, that fallen trees from many years ago are still commercially valuable.

Merbau is still a popular decking timber and has a long life in above ground applications, but it is logged in an unsustainable manner and is too hard and brittle for general use.

Ironbark is a good choice where available. It is harvested from both old-growth and plantation in Australia and is highly resistant to rot and termites. It is most commonly used for fence posts and house stumps.

Eastern red cedar (*Juniperus virginiana*) and black locust (*Robinia pseudoacacia*) have long been used for rot-resistant fence posts and rails in eastern United States, with the black locust also planted in modern times in Europe. Coast redwood is commonly used for similar applications in the western United States.

Tōtara and puriri were used extensively in New Zealand during the European colonial era when native forests were "mined", even as fence posts of which many are still operating. Tōtara was used by the Māori to build large *waka* (canoes). Today, they are specialty timbers as a result of their scarcity, although lower grade stocks are sold for landscaping use.

Kauri is a superb timber for building the hulls and decks of boats. It too is now a specialty timber and ancient logs (in excess of 3,000 years) that have been mined from swamps are used by wood turners and furniture makers.

The natural durability or rot and insect resistance of wood species is always based on the heartwood (or "truewood"). The sapwood of all timber species should be considered to be non-durable without preservative treatment.

Tung Oil

Tung oil has been used for hundreds of years in China, where it was used as a preservative for wood ships. The oil penetrates the wood, and then hardens to form an impermeable hydrophobic layer up to 5 mm into the wood. As a preservative it is effective for exterior work above and below ground, but the thin layer makes it less useful in practice. It is not available as a pressure treatment.

Heat Treatments

By going beyond kiln drying wood, heat treatment may make timber more durable. By heating timber to a certain temperature, it may be possible to make the wood fibre less appetizing to insects.

Heat treatment can also improve the properties of the wood with respect to water, with lower equilibrium moisture, less moisture deformation, and weather resistance. It is weather-resistant enough to be used unprotected, in facades or in kitchen tables, where wetting is expected.

There are four similar heat treatments — Westwood, developed in the United States; Retiwood, developed in France; Thermowood, developed in Finland by VTT; and Platowood, developed in The Netherlands. These processes autoclave the treated wood, subjecting it to pressure and heat, along with nitrogen or water vapour to control drying in a staged treatment process ranging from 24 to 48 hours at temperatures of 180 °C to 230 °C depending on timber species. These processes increase the durability, dimensional stability and hardness of the treated wood by at least one class; however, the treated wood is darkened in colour, and there are changes in certain mechanical characteristics: Specifically, the modulus of elasticity is increased to 10%, and the modulus of rupture is diminished by 5% to 20%; thus, the treated wood requires drilling for nailing to avoid splitting the wood. Certain of these processes cause less of an impact than others in their the mechanical effects upon the treated wood. Wood treated with this process is often used for cladding or siding, flooring, furniture and windows.

For the control of pests that may be harbored in wood packaging material (i.e. crates and pallets), the ISPM 15 requires heat treatment of wood to 56 °C for 30 minutes to receive the HT stamp. This is typically required to ensure the killing of the pine wilt nematode and other kinds of wood pests that could be transported internationally.

Mud Treatment

Wood and bamboo can be buried in mud to help protect them from insects and decay. This practice is used widely in Vietnam to build farm houses consisting of a wooden structural frame, a bamboo roof frame and bamboo with mud mixed with rice hay for the walls. While wood in contact with soil will generally decompose more quickly than wood not in contact with it, it is possible that the predominantly clay soils prevalent in Vietnam provide a degree of mechanical protection against insect attack, which compensates for the accelerated rate of decay.

Also, since wood is only subject to bacterial decay under specific temperature and moisture content ranges, submerging it in water-saturated mud can retard decay, by saturating the wood's internal cells beyond their moisture decay range.

Application Processes

Introduction and History

Probably the first attempts made to protect wood from decay and insect attack consisted of brushing or rubbing preservatives onto the surfaces of the treated wood. Through trial and error the most effective preservatives and application processes were slowly determined. In the Industrial Revolution, demands for such things as telegraph poles and railroad ties (UK: railway sleepers) helped to fuel an explosion of new techniques that emerged in the early 19th century. The sharpest rise in inventions took place between 1830 and 1840, when Bethell, Boucherie, Burnett and Kyan were making wood-preserving history. Since then, numerous processes have been introduced or existing processes improved. The goal of modern-day wood preservation is to ensure a deep, uniform penetration with reasonable cost, without endangering the environment. The most widespread application processes today are those using artificial pressure through which many woods are being effectively treated, but several species (such as spruce, Douglas-fir, larch, hemlock and fir) are very resistant to impregnation. With the use of incising, the treatment of these woods has been somewhat successful but with a higher cost and not always satisfactory results. One can divide the wood-preserving methods roughly into either non-pressure processes or pressure processes.

Non-pressure Processes

There are numerous non-pressure processes of treating wood which vary primarily in their procedure. The most common of these treatments involve the application of the preservative by means of brushing or spraying, dipping, soaking, steeping or by means of hot and cold bath. There is also a variety of additional methods involving charring, applying preservatives in bored holes, diffusion processes and sap displacement.

Brush and Spray Treatments

Brushing preservatives is a long-practised method and often used in today's carpentry workshops. Technological developments mean it is also possible to spray preservative over the surface of the timber. Some of the liquid is drawn into the wood as the result of capillary action before the spray runs off or evaporates, but unless puddling occurs penetration is limited and may not be suitable for long-term weathering. By using the spray method, coal-tar creosote, oil-borne solutions and water-borne salts (to some extent) can also be applied. A thorough brush or spray treatment with coal-tar creosote can add 1 to 3 years to the lifespan of poles or posts. Two or more coats provide better protection than one, but the successive coats should not be applied until the prior coat has dried or soaked into the wood. The wood should be seasoned before treatment.

Dipping

Dipping consists of simply immersing the wood in a bath of creosote or other preservative for a few seconds or minutes. Similar penetrations to that of brushing and spraying processes are achieved. It has the advantage of minimizing hand labor. It requires more equipment and larger quantities of preservative and is not adequate for treating small lots of timber. Usually the dipping process is useful in the treatment of window sashes and doors. Treatment with copper salt preservatives is no longer allowed with this method.

Steeping

In this process the wood is submerged in a tank of water-preservative mix, and allowed to soak for a longer period of time (several days to weeks). This process was developed in the 19th century by John Kyan. The depth and retention achieved depends on factors such as species, wood moisture, preservative and soak duration. The majority of the absorption takes place during the first two or three days, but will continue at a slower pace for an indefinite period. As a result, the longer the wood can be left in the solution, the better treatment it will receive. When treating seasoned timber, both the water and the preservative salt soak into the wood, making it necessary to season the wood a second time. Posts and poles can be treated directly on endangered areas, but should be treated at least 30 cm (0.98 ft) above the future ground level.

The depth obtained during regular steeping periods varies from 5 to 10 mm (0.20 to 0.39 in) up to 30 mm (1.2 in) by sap pine. Due to the low absorption, solution strength should be somewhat stronger than that in pressure processes, around 5% for seasoned timber and 10% for green timber (because the concentration slowly decreases as the chemicals diffuse into the wood). The solution strength should be controlled continually and, if necessary, be corrected with the salt additive. After the timber is removed from the treatment tank, the chemical will continue to spread within the wood if it has sufficient moisture content. The wood should be weighed down and piled so that the solution can reach all surfaces. (Sawed materials stickers should be placed between every board layer.) This process finds minimal use despite its former popularity in continental Europe and Great Britain.

Kyanizing

Named after John Howard Kyan, who patented this process in England in 1833, Kyanizing consists of steeping wood in a 0.67% mercuric chloride preservative solution.

Gedrian's Bath

Patented by Charles A. Seely, this process achieves treatment by immersing seasoned wood in successive baths of hot and cold preservatives. During the hot baths, the air expands in the timbers. When the timbers are changed to the cold bath (the preservative can also be changed) a partial vacuum is created within the lumen of the cells, causing the preservative to be drawn into the wood. Some penetration occurs during the hot baths, but most of it takes place during the cold baths. This cycle is repeated with a significant time reduction compared to other steeping processes. Each bath may last 4 to 8 hours or in some cases longer. The temperature of the preservative in the hot bath should be between 60 to 110 °C (140 to 230 °F) and 30 to 40 °C (86 to 104 °F) in the cold bath (depending on preservative and tree species). The average penetration depths achieved with this process ranges from 30 to 50 mm (1.2 to 2.0 in). Both preservative oils and water-soluble salts can be used with this treatment. Due to the longer treatment periods, this method finds little use in the commercial wood preservation industry today.

Preservative Precipitation

As explained in Uhlig's Corrosion Handbook, this process involves two or more chemical baths that undergo a reaction with the cells of the wood, and result in the precipitation of preservative into the wood cells. Two chemicals commonly employed in this process are Copper Ethanolamine, and Sodium Dimethyldithiocarbamate, which reacts to precipitate Copper Dimetyldithiocarbamate. The precipitated preservative is very resistant to leeching. Since its use in the mid 1990s, it has been discontinued in the United States of America, however it never saw commercialization in Canada.

Pressure Processes

Late 19th century pressure treatment.

Pressure processes are the most permanent method around today in preserving timber life. Pressure processes are those in which the treatment is carried out in closed cylinders with applied pressure and/or vacuum. These processes have a number of advantages over the non-pressure methods. In most cases, a deeper and more uniform penetration and a higher absorption of preservative is achieved. Another advantage is that the treating conditions can be controlled so that retention and penetration can be varied. These pressure processes can be adapted to large-scale production. The high initial costs for equipment and the energy costs are the biggest disadvantages. These treatment methods are used to protect ties, poles and structural timbers and find use throughout the world today. The various pressure processes that are used today differ in details, but the general method is in all cases the same. The treatment is carried out in cylinders. The timbers are loaded onto special tram cars, so called *buggies* or *bogies*, and into the cylinder. These cylinders are then set under pressure often with the addition of higher temperature. As final treatment, a vacuum is frequently used to extract excess preservatives. These cycles can be repeated to achieve better penetration.

LOSP treatments often use a vacuum impregnation process. This is possible because of the lower viscosity of the white-spirit carrier used.

Full-cell Process

In the full-cell process, the intent is to keep as much of the liquid absorbed into the wood during the pressure period as possible, thus leaving the maximum concentration of preservatives in the treated area. Usually, water solutions of preservative salts are employed with this process, but it is also possible to impregnate wood with oil. The desired retention is achieved by changing the strength of the solution. William Burnett patented this development in 1838 of full-cell impregnation with water solutions. The patent covered the use of zinc chloride on water basis, also known as *Burnettizing*. A full-cell process with oil was patented in 1838 by John Bethell. His patent described the injection of tar and oils into wood by applying pressure in closed cylinders. This process is still used today with some improvements.

Fluctuation Pressure Process

Contrary to the static full-cell and empty-cell processes, the fluctuation process is a dynamic process. By this process the pressure inside the impregnation cylinder changes between pressure and vacuum within a few seconds. There have been inconsistent claims that through this process it is possible to reverse the pit closure by spruce. However, the best results that have been achieved with this process by spruce do not exceed a penetration deeper than 10 mm (0.39 in). Specialized equipment is necessary and therefore higher investment costs are incurred.

Boucherie Process

Developed by Dr. Boucherie of France in 1838, this approach consisted of attaching a bag or container of preservative solution to a standing or a freshly cut tree with bark, branches, and leaves still attached, thereby injecting the liquid into the sap stream. Through transpiration of moisture from the leaves the preservative is drawn upward through the sapwood of the tree trunk.

The modified Boucherie process consists of placing freshly cut, unpeeled timbers onto declining skids, with the stump slightly elevated, then fastening watertight covering caps or boring a number of holes into the ends, and inserting a solution of copper sulfate or other waterborne preservative into the caps or holes from an elevated container. Preservative oils tend to not penetrate satisfactorily by this method. The hydrostatic pressure of the liquid forces the preservative lengthwise into and through the sapwood, thus pushing the sap out of the other end of the timber. After a few days, the sapwood is completely impregnated; unfortunately little or no penetration takes place in the heartwood. Only green wood can be treated in this manner. This process has found considerable usage to impregnate poles and also larger trees in Europe and North America, and has experienced a revival of usage to impregnate bamboo in countries such as Costa Rica, Bangladesh, India and the state of Hawaii.

High-pressure Sap Displacement System

Developed in the Philippines, this method (abbreviated HPSD) consists of a cylinder pressure cap made from a 3 mm thick mild steel plate secured with 8 sets of bolts, a 2-HP diesel engine, and a pressure regulator with 1.4–14 kg/m^2 capacity. The cap is placed over the stump of a pole, tree or bamboo and the preservative is forced into the wood with pressure from the engine.

Incising

First tested and patented by Kolossvary, Haltenberger, and Berdenich of Austria in 1911 and 1912 (U.S. pats. 1,012,207 and 1,018,624) with several improvements from O. P. M. Goss, D. W. Edwards and J. H. Mansfield among others, this process consists of making shallow, slit-like holes in the surfaces of material to be treated, so that deeper and more uniform penetration of preventative may be obtained. The term *incising* or perforating comes from the Latin *incidere*, a compound of *in* and *caedere* (to cut). Incisions made in sawed material usually are parallel with the grain of the wood. This process is common in North America (since the 1950s), where Douglas-fir products and pole butts of various species are prepared before treatment. It is most useful for woods that are resistant to side penetration, but allow preservative transport along the grain. In the region in which it is produced, it is common practice to incise all sawed Douglas-fir 3 in (76 mm) or more in thickness before treatment.

Unfortunately, the impregnation of spruce, the most important structural timber in large areas in Europe, has shown that unsatisfactory treatment depths have been achieved with impregnation. The maximum penetration of 2 mm (0.079 in) is not sufficient to protect wood in weathered positions. The present-day incising machines consist essentially of four revolving drums fitted with teeth or needles or with lasers that burn the incisions into the wood. Preservatives can be spread along the grain up to 20 mm (0.79 in) in radial and up to 2 mm (0.079 in) in tangential and radial direction.

In North America, where smaller timber dimensions are common, incision depths of 4 to 6 mm (0.16 to 0.24 in) have become standard. In Europe, where larger dimensions are widespread, incision depths of 10 to 12 mm (0.39 to 0.47 in) are necessary. The incisions are visible and often considered to be wood error. Incisions by laser are significantly smaller than those of spokes or needles. The costs for each process type are approximately for spoke/conventional all-round incising €0.50/m^2, by laser incising €3.60/m^2 and by needle incision €1.00/m^2. (Figures originate from the year 1998 and may vary from present day prices.)

Microwaving

An alternative increases the permeability of timber using microwave technology. There is some concern that this method may adversely affect the structural performance of the material. Research in this area has been conducted by the Cooperative Research Centre at the University of Melbourne, Australia.

Charring

Charring of timber results in surfaces which are fire-resistant, insect-resistant and proof against weathering. Wood surfaces are ignited using a hand-held burner or moved slowly across a fire. The charred surface is then cleaned using a steel brush to remove loose bits and to expose the grain. Oil or varnish may be applied if required.

Timber Recycling

Demolishers pulling timber from an old wool store in Sydney, which will later be re-used for timber flooring.

Timber recycling or wood recycling is the process of turning waste timber into usable products. Recycling timber is a practice that was popularized in the early 1990s as issues such as deforestation and climate change prompted both timber suppliers and consumers to turn to a more sustainable timber source. Recycling timber is the environmentally friendliest form of timber production and is very common in countries such as Australia and New Zealand where supplies of old wooden structures are plentiful. Timber can be chipped down into wood chips which can be used to power homes or power plants.

Benefits

Recycling timber has become popular due to its image as an environmentally friendly product. Common belief among consumers is that by purchasing recycled wood, the demand for "green timber" will fall and ultimately benefit the environment. Greenpeace also view recycled timber as an environmentally friendly product, citing it as the most preferable timber source on their website. The arrival of recycled timber as a construction product has been important in both raising industry and consumer awareness towards deforestation and promoting timber mills to adopt more environmentally friendly practices.

Example of recycled timber as a final product.

Drawbacks

Some hurdles facing the widespread adoption of recycled timber: sometimes the ends of wall studs need to be trimmed off to stop decay and cracking, thus resulting in a shorter piece of wood; this trimming may result in pieces of wood that do not meet building codes. Though the price may be less than for new wood, the process of selecting usable pieces of salvaged wood, pulling out nails, and refinishing for a new use can be laborious and time-consuming. Demolition must happen in such a way as to preserve as much of the timber as possible in a building, which means more time spent dismantling a building rather than just tearing it down quickly. The trade in recycled timber is not well-established everywhere, so a reliable supply of usable wood may be hard to come by for builders. There may be a stigma associated with using "used" or "cheap" wood that is perceived to be of not as high quality as "new" wood. Not all pieces of wood in a dismantled building will fit in a new building, and it may be cheaper and easier, from a design and labor perspective, to simply get new wood (ex: wood from a 6-foot (1.8 m) deck being used in a 7-foot (2.1 m) deck). Of course, none of these issues are insurmountable, and they are issues of convenience and logistics rather than structural integrity, but many builders find it easier and less time-consuming to simply get new wood in standard uniform sizes.

Recycling Timber

Recycled timber most commonly comes from old buildings, bridges and wharfs, where it is carefully stripped out and put aside by demolishers. At the same time any usable dimension stone is set aside for reuse. The demolishers then sell the salvaged timber to merchants who then re-mill the timber by manually scanning it with a metal detector, which allows the timber to be de-nailed and sawn to size. Once re-milled the timber is commonly sold to consumers in the form of timber flooring, beams and decking.

Recycled timber salvaged from a demolished building in Trat, Thailand, is applied to make a new roof.

Examples

Use of recycling timber is not new. As early as 1948, the 100 metres tall tower of Golm transmitter near Potsdam, Germany was built from recycled timber. It stood for 31 years. It was common to reuse wood of dismantled radio towers in the 1930s in Germany, e.g. the former tower of Koblenz radio transmitter was built of wood from a dismantled tower, which carried a T-antenna at Transmitter Muehlacker. The upper parts of the 157-metre-tall wood tower of Ismaning radio transmitter, which stood for 49 years, were built of wood from a smaller radio tower dismantled in 1934.

Non-timber Forest Product

Non-timber forest products (NTFPs), also special, non-wood, minor, alternative and secondary forest products, are useful substances, materials and/or commodities obtained from forests which do not require harvesting (logging) trees. They include game animals, fur-bearers, nuts, seeds, berries, mushrooms, oils, foliage, medicinal plants, peat, fuelwood, and forage.

Mahua seeds

Tendu patta (leaf) collection.

Research on NTFPs have focused on their commodifiability for rural incomes and markets, as an expression of traditional knowledge or as a livelihood option for rural household needs, and, as a key component of sustainable forest management and conservation strategies. All research promote forest products as valuable commodities and tools that can promote the conservation of forests.

Definitions

There is a wide variety of NTFPs, including mushrooms, huckleberries, ferns, transplants, seed cones, piñon seeds, tree nuts, moss, maple syrup, cork, cinnamon, rubber, tree oils and resins, and ginseng. The United Kingdom's Forestry Commission defines NTFPs as "any biological resources found in woodlands except timber," and Forest Harvest, part of the Reforesting Scotland project, defines them as "materials supplied by woodlands - except the conventional harvest of timber." These definitions include wild and managed game, fish and insects. NTFPs are commonly grouped into categories such as floral greens, decoratives, medicinal plants, foods, flavors and fragrances, fibers, and saps and resins.

Other terms similar to NTFPs include *special, non-wood, minor, alternative* and *secondary forest products*. NTFPs in particular highlight forest products which are of value to local people and communities but have been overlooked in the wake of forest management priorities (for example, timber production and animal forage). In recent decades, interest has grown in using NTFPs as alternatives or supplements to forest management practices. In some forest types, under the right political and social conditions, forests can be managed to increase NTFP diversity and, consequently, to increase biodiversity and potentially economic diversity. Black truffle cultivation in the mediterranean area is a good example of a high profitability when well managed.

Uses

The harvest of NTFPs remains widespread throughout the world. People from a wide range of socio-economic, geographical and cultural contexts harvest NTFPs for a number of purposes, including but not limited to: household subsistence, maintenance of cultural and familial traditions, spiritual fulfillment as well as physical and emotional well-being, scientific learning and income. Other terms synonymous with harvesting include *wild-crafting, gathering, collecting* and *foraging*. NTFPs also serve as raw materials for industries ranging from large-scale floral greens suppliers and pharmaceutical companies to micro-enterprises centred upon a wide variety of activities (such as basket-making, woodcarving and the harvest and processing of various medicinal plants).

Economic Importance

It is difficult to estimate the contribution of NTFPs to national or regional economies as there is a lack of broad-based systems for tracking the combined value of the hundreds of products that make up various NTFP industries. One exception to this is the maple syrup industry, which in 2002 in the US alone yielded 1.4 million US gallons (5,300 m^3) worth USD$38.3 million. In temperate forests such as in the US, wild edible mushrooms such as matsutake, medicinal plants such as ginseng, and floral greens such as salal and *sword fern* are multimillion-dollar industries. While these high-value species may attract the most attention, a diversity of NTFPs can be found in most forests of the world.

In tropical forests, for example, NTFPs can be an important source of income that can supplement farming and/or other activities. A value-analysis of the Amazon rainforest in Peru found that exploitation of NTFPs could yield higher net revenue per hectare than would timber harvest of the same area, while still conserving vital ecological services. Their economic, cultural and ecological value, when considered in aggregate, makes managing NTFPs an important component of sustainable forest management and the conservation of biological and cultural diversity.

Research

Research on NTFPs have focused on three perspectives: NTFPs as a commodity with a focus on rural incomes and markets, as an expression of traditional knowledge or as a livelihood option for rural household needs, and, finally, as a key component of sustainable forest management and conservation strategies. These perspectives promote forest products as valuable commodities and important tools that can promote the conservation of forests. In some contexts, the gathering and use of NTFPs can be a mechanism for poverty alleviation and local development.

Bụt gave to him four magic words to attach the many nodes of bamboo: *Khắc nhập, khắc xuất*, which means "joined together immediately, fell apart immediately". The triumphant farmer returned to the landlord and demanded his daughter. Curious to see such a long bamboo, the landlord was magically joined to the bamboo when he touched it, as the young farmer said the first two magic words. The story ends with the happy marriage of the farmer and the landlord's daughter after the landlord agreed to the marriage and asked to be separated from the bamboo.

In a Chinese legend, the Emperor Yao gave two of his daughters to the future Emperor Shun as a test for his potential to rule. Shun passed the test of being able to run his household with the two emperor's daughters as wives, and thus Yao made Shun his successor, bypassing his unworthy son.

Later, Shun drowned in the Xiang River. The tears his two bereaved wives let fall upon the bamboos growing there explains the origin of spotted bamboo. The two women later became goddesses.

According to Maranaw

- Buntong or Bentong (e.i Palaw ah Bentong that means Mountain of Bamboos).
- Tringen *Tring.*

References

- Federal Institute for Risk Assessment (Germany). "Migration of mineral oil from packaging materials to foodstuffs (BfR Opinion No. 008/2010 of 2009-12-09)" (PDF). Retrieved 2010-04-21
- Bundesinstitut für Risikobewertung. "Übergänge von Mineralöl aus Verpackungsmaterialien auf Lebensmittel" (PDF). Retrieved 2010-04-21
- "Wood waste: A short review of recent research" (PDF). Department for Environment, Food and Rural Affairs. July 2012. Retrieved 5 July 2016
- "Factors affecting distribution of borate to protect building envelope components from biodegradation" (PDF). Tspace.library.utoronto.ca. Retrieved 2015-08-27
- "Forest Research - Social, cultural and economic values of contemporary non-timber forest products: Wild Harvests". Forestry.gov.uk. Retrieved 2013-11-21
- Iggesund Paperboard (2008). "Product Catalogue: General Technical Information". Archived from the original on July 13, 2011. Retrieved 2009-07-18
- Metso Corporation's press release (2011). "Metso to supply coated board line to International Paper & Sun Cartonboard in China". Retrieved 2011-11-04
- Soroka, W (2008). Illustrated Glossary of Packaging Terms. Institute of Packaging Professionals. p. 81. ISBN 1-930268-27-0
- MacDonald, Nancy (2013). Woodworking (2nd ed.). Cengage Learning. pp. 782–784. ISBN 9781285700502. Retrieved 12 November 2015

Permissions

All chapters in this book are published with permission under the Creative Commons Attribution Share Alike License or equivalent. Every chapter published in this book has been scrutinized by our experts. Their significance has been extensively debated. The topics covered herein carry significant information for a comprehensive understanding. They may even be implemented as practical applications or may be referred to as a beginning point for further studies.

We would like to thank the editorial team for lending their expertise to make the book truly unique. They have played a crucial role in the development of this book. Without their invaluable contributions this book wouldn't have been possible. They have made vital efforts to compile up to date information on the varied aspects of this subject to make this book a valuable addition to the collection of many professionals and students.

This book was conceptualized with the vision of imparting up-to-date and integrated information in this field. To ensure the same, a matchless editorial board was set up. Every individual on the board went through rigorous rounds of assessment to prove their worth. After which they invested a large part of their time researching and compiling the most relevant data for our readers.

The editorial board has been involved in producing this book since its inception. They have spent rigorous hours researching and exploring the diverse topics which have resulted in the successful publishing of this book. They have passed on their knowledge of decades through this book. To expedite this challenging task, the publisher supported the team at every step. A small team of assistant editors was also appointed to further simplify the editing procedure and attain best results for the readers.

Apart from the editorial board, the designing team has also invested a significant amount of their time in understanding the subject and creating the most relevant covers. They scrutinized every image to scout for the most suitable representation of the subject and create an appropriate cover for the book.

The publishing team has been an ardent support to the editorial, designing and production team. Their endless efforts to recruit the best for this project, has resulted in the accomplishment of this book. They are a veteran in the field of academics and their pool of knowledge is as vast as their experience in printing. Their expertise and guidance has proved useful at every step. Their uncompromising quality standards have made this book an exceptional effort. Their encouragement from time to time has been an inspiration for everyone.

The publisher and the editorial board hope that this book will prove to be a valuable piece of knowledge for students, practitioners and scholars across the globe.

Index

A

Afforestation, 6-7, 28-29, 51, 56, 63-64, 76-80, 89-90, 94, 106, 141, 157, 178-179, 187-188

Agroforestry, 1, 16-26, 52, 188

Alley Cropping, 17-18, 20-22, 24

Analog Forestry, 2

B

Biodiversity, 10, 26, 33-34, 40, 45, 48, 52-53, 57, 76, 79-80, 84-85, 90, 92, 97, 101, 104, 109, 111-112, 141, 159, 161, 164-165, 167, 170, 176, 182, 187-189, 229

Biofuels, 32, 110-111, 130-132

Biomass, 2, 7, 11, 19, 26-27, 32, 44, 82, 84, 86-87, 96, 107-109, 112-114, 116, 118, 123-124, 126, 130-132, 134, 141, 158, 177, 187

Biome, 189

Biosequestration, 89, 102-104, 107-112, 114, 160

Buffer Strips, 37

C

Carbon Capture, 76, 111-114, 118-119, 130, 142-143

Carbon Cycle, 10-11, 90, 114-116

Carbon Emissions, 53, 80, 91-92, 104, 112, 122, 131, 137, 139, 141, 143, 149, 152, 154, 157-158

Carbon Sequestration, 11, 26, 76, 112-120, 143, 147, 158, 177

Carbon Sink, 9, 11, 28, 33-34, 58, 104-105, 114-115, 140

Clearcutting, 28, 82, 86

Climate Change Mitigation, 90, 103, 110, 124-126, 128-130, 144, 153-155

Cloud Forests, 57

Coarse Woody Debris, 50

Community Forestry, 2, 53, 100, 179

Conifer Forests, 51

Coniferous Forest, 174

Coniferous Forests, 60-61

Copper Azole, 211-213

Copper Naphthenate, 213

Coppice, 3, 27, 33-35, 44, 76

D

Deciduous Forest, 61

Deciduous Forests, 60

Deforestation, 62, 77-78, 80, 89, 91-94, 103-107, 111-112, 141, 159-175, 177-182, 184, 187, 189, 196, 226

Douglas Fir, 61, 89

E

Ecoforestry, 39-41

Energy Crop, 2

Energy Forestry, 2, 26-27, 35

Eucalyptus, 11, 27, 90, 95, 219

Evapotranspiration, 163

Evergreen Forest, 193

F

Fiberboard, 198-199, 203-205

Food Web, 140

Forage, 17, 19, 22, 31, 44, 51, 228-229

Forest Conservation, 2, 61-62, 99, 161

Forest Cover, 6, 11, 48, 62, 64, 76-78, 80, 90-93, 106, 112, 141, 163-167, 173, 175-176, 187-188

Forest Crop, 2

Forest Degradation, 105, 111-112, 162-163, 175, 187

Forest Dieback, 159, 189-190, 192-193

Forest Floor, 50, 81, 87-89

Forest Management, 1-2, 6-7, 9-10, 37, 43-44, 46, 48-49, 51, 64-72, 75, 94, 97, 102, 106, 112, 160, 178-179, 182, 184, 189, 229-230

Forest Stands, 44, 46, 48, 56, 82, 90, 159

Forest structure, 42, 44-46, 50, 188

G

Gap Dynamics, 76, 81-82, 85, 157

Geoengineering, 113, 117, 142-143, 158

H

Habitat Loss, 173

Heat Pump, 136-137, 139

I

Illegal Logging, 159, 165-166, 173, 181-184, 186

K

Kyoto Protocol, 105-106, 112, 144, 150, 152, 154

L

Life Cycle Assessment, 7-8, 11-12, 14-15, 133

M

Mangrove Forests, 58, 112, 174, 179

Mulch, 17, 20, 30

N

Natural Landscapes, 47

Natural Regeneration, 40, 83, 90, 159

O

Organic Matter, 18, 20-21, 26, 56, 90, 108

P

Paperboard, 194, 198-203, 231

Perennials, 16, 25, 86

Precipitation, 56-57, 59, 61-62, 86-87, 163, 223

Priority Species, 41, 44, 48-49

Pruning, 96

R

Reforestation, 7, 56, 63, 76-77, 89-92, 94-95, 102, 104, 106-107, 112, 114-115, 125, 137, 141, 143, 159, 178, 187-188

Rotation Plantation, 27, 29, 32

S

Shifting Cultivation, 17, 22, 24, 64

Short Rotation Plantation, 27, 29, 32

Silviculture, 2, 21, 43-46, 48-51, 69, 177

Social Forestry, 2, 51-55

Soil Fertility, 6, 18, 21, 26, 108

Sustainable Forest, 2, 6, 9, 67, 106, 178, 189, 229-230

Sustainable Forestry, 36-39

T

Taiga, 56, 61

Taiga Forests, 61

Transpiration, 56, 163, 168, 224

Tree Canopy, 100-101

Tree Species, 1, 16-17, 19-20, 27, 30-31, 43, 46, 50, 56, 83-85, 101, 191, 205, 219, 223

Tropical Rainforest, 60, 105, 165, 172

U

Urban Forestry, 2, 96, 98-102, 115

W

Wildlife Forestry, 41, 43, 46-49

Wildlife Habitat, 10, 36, 42, 44-45

Wildlife Management, 48

Wood Preservation, 194, 207-211, 215, 221, 223

Wood Veneer, 204-207

Woodland, 28, 41, 80, 95, 98, 101, 171, 178

Woodworking, 194-197, 204, 231

www.ingramcontent.com/pod-product-compliance
Lightning Source LLC
LaVergne TN
LVHW080728210926
848485LV00003B/535
* 9 7 8 1 6 3 9 8 7 3 3 1 9 *